ELECTRONIC S TECHNICAL, AND MEDICAL JOURNAL PUBLISHING AND ITS IMPLICATIONS

M000092201

REPORT OF A SYMPOSIUM

Committee on Electronic Scientific, Technical, and
Medical Journal Publishing

Committee on Science, Engineering, and Public Policy
Policy and Global Affairs Division

THE NATIONAL ACADEMIES

THE NATIONAL ACADEMIES PRESS
Washington, DC
www.nap.edu

THE NATIONAL ACADEMIES PRESS • 500 Fifth Street, NW • Washington, DC 20001

NOTICE: The project that is the subject of this report was approved by the Governing Board of the National Research Council, whose members are drawn from the councils of the National Academy of Sciences, the National Academy of Engineering, and the Institute of Medicine. The members of the committee responsible for the report were chosen for their special competences and with regard for appropriate balance.

Support for this project was provided by the National Academy of Sciences through an unnumbered internal grant. Any opinions, findings, conclusions, or recommendations expressed in this publication are those of the author(s) and symposium speakers and do not necessarily reflect the views of the organizations or agencies that provided support for the project.

International Standard Book Number 0-309-09161-6 (Book)
International Standard Book Number 0-309-53061-X (PDF)

Committee on Science, Engineering, and Public Policy, 500 Fifth Street, NW, Washington, DC 20001; 202-334-2424; Internet, http://www7.nationalacademies.org/cosepup/

Additional copies of this report are available from the National Academies Press, 500 Fifth Street, NW, Lockbox 285, Washington, DC 20055; (800) 624-6242 or (202) 334-3313 (in the Washington metropolitan area); Internet, http://www.nap.edu

THE NATIONAL ACADEMIES
Advisers to the Nation on Science, Engineering, and Medicine

The **National Academy of Sciences** is a private, nonprofit, self-perpetuating society of distinguished scholars engaged in scientific and engineering research, dedicated to the furtherance of science and technology and to their use for the general welfare. Upon the authority of the charter granted to it by the Congress in 1863, the Academy has a mandate that requires it to advise the federal government on scientific and technical matters. Dr. Bruce M. Alberts is president of the National Academy of Sciences.

The **National Academy of Engineering** was established in 1964, under the charter of the National Academy of Sciences, as a parallel organization of outstanding engineers. It is autonomous in its administration and in the selection of its members, sharing with the National Academy of Sciences the responsibility for advising the federal government. The National Academy of Engineering also sponsors engineering programs aimed at meeting national needs, encourages education and research, and recognizes the superior achievements of engineers. Dr. Wm. A. Wulf is president of the National Academy of Engineering.

The **Institute of Medicine** was established in 1970 by the National Academy of Sciences to secure the services of eminent members of appropriate professions in the examination of policy matters pertaining to the health of the public. The Institute acts under the responsibility given to the National Academy of Sciences by its congressional charter to be an adviser to the federal government and, upon its own initiative, to identify issues of medical care, research, and education. Dr. Harvey V. Fineberg is president of the Institute of Medicine.

The **National Research Council** was organized by the National Academy of Sciences in 1916 to associate the broad community of science and technology with the Academy's purposes of furthering knowledge and advising the federal government. Functioning in accordance with general policies determined by the Academy, the Council has become the principal operating agency of both the National Academy of Sciences and the National Academy of Engineering in providing services to the government, the public, and the scientific and engineering communities. The Council is administered jointly by both Academies and the Institute of Medicine. Dr. Bruce M. Alberts and Dr. Wm. A. Wulf are chair and vice chair, respectively, of the National Research Council.

www.national-academies.org

Preface

The use of the Internet and other digital information technologies by the scientific, technical, and medical (STM) research community in the United States and most other countries has transformed many aspects of the research and publishing process. The new technologies have created fundamental changes in the production, management, dissemination, and use of all types of information. It is now possible to communicate research results much more quickly, broadly, and openly than was possible through traditional print publications in the past. Researchers are now able to make available independently their data and articles online, where the information may be easily found, browsed, annotated, critiqued, downloaded, and freely shared. This is resulting in significant changes to the linear path of writing, refereeing, and reviewing of publications as all these functions can be performed concurrently. Most STM publishers also now publish electronic versions of their journals, some exclusively so. The technological developments and resulting changes to the sociology of science are creating both opportunities and challenges for the effective management of scientific communication generally, and STM publishing more specifically.

Because of the far-reaching implications of these developments, the National Academy of Sciences Council's Committee on Publications recommended that the Council commission a study of the factors involved in the changing mechanisms for access to STM information in the scholarly publications and the various technical, legal, policy, and economic issues that they raise. The committee indicated that it is imperative for the National Academies to address, in particular, the increasing concerns about the implications of various models for access to STM publications for the scientific community.

As a result, the Committee on Science, Engineering, and Public Policy was asked to appoint a committee to oversee the planning for the *Symposium on Electronic Scientific, Technical, and Medical Journal Publishing and Its Implications,* which was held May 19-20, 2003, at the National Academy of Sciences in Washington, D.C. The symposium brought together experts in STM publishing, both producers and users of these publications, to: (1) identify the recent technical changes in publishing, and other factors, that influence the decisions of journal publishers to produce journals electronically; (2) identify the needs of the scientific, engineering, and medical community as users of journals, whether electronic or printed; (3) discuss the responses of not-for-profit and commercial STM publishers and of other stakeholders in the STM community to the opportunities and challenges posed by the shift to electronic publishing; and (4) examine the spectrum of proposals that has been put forth to respond to the needs of users as the publishing industry shifts to electronic information production and dissemination.

The symposium was divided into six sessions, each introduced by opening comments from a moderator, followed by several invited presentations. Session 1 examined the costs involved with the publication of STM journals while Session 2 looked at the related publication business models. Session 3 explored the legal issues in the production and dissemination of these journals. Sessions 4 and 5 looked toward the future and examined, respectively, what is publication in the future and what constitutes a publication in the digital environment. The final session provided several commentaries on the presentations and discussions that took place during the symposium.

The proceedings of the symposium were taped and transcribed, and served as the basis for this symposium report. The formal *Proceedings* of the symposium are available online via the National Academies Press. The National Academies hosted a live audio webcast of the symposium to reach a broad audience and receive additional input. The webcast, along with the edited *Proceedings* of the symposium, can be found on the symposium Web site at http://www7.nationalacademies.org/cosepup/E-Publishing.html.

This report is based on excerpts from the symposium *Proceedings* that the committee found particularly useful to highlight. It summarizes the views of the symposium participants but does not contain any consensus findings, conclusions, or recommendations of the committee itself. A footnote in each major section of the report identifies the individuals responsible for the views presented there. In addition, Chapters 3-7 each have a final section entitled "Issues Raised in the Discussion" that summarizes portions of the general discussion of the expert invited panel speakers among themselves and with the audience. Because of the large number of speakers providing comments in those discussions, we have not attributed each point to specific individuals. However, the source of each point may be found in the edited online *Proceedings*.

The report does not cover all the issues that a more comprehensive study might consider, since it is limited only to the topics discussed during the symposium. Several important issues were identified in subsequent discussions, including: the relationship of journal size (number of "pages," articles, issues) to total annual cost; the "first copy" costs and the relationship of those costs to income sources such as subscriptions or payments from aggregators; and federal policies on the publication of articles in STM journals by recipients of federal research support.

Nor was it possible to involve representatives of all types of STM journals or of all functions in the broader process of scientific communication. The comments and suggestions made by the various participants cannot be generalized easily across the entire spectrum of publications in the STM journal enterprise. For example, significant differences exist between for-profit and not-for-profit journals, between those that represent professional societies and those that do not, between journals that derive revenues from advertisements and those that do not, and between clinical and basic science journals. With regard to the latter, for instance, editors of clinical journals caution that they have a special responsibility in quality control and review. Whereas readers of basic science journals are generally scientists who can critique the articles they read, those who turn to clinical journals for current knowledge frequently may not be experts in the research disciplines covered and have come to depend on the editorial process to assure the accuracy and validity of the papers that they read. The quality and safety of patient care may thus be tied rather directly to the quality of the editorial process in clinical journals.

Despite these acknowledged limitations of the symposium discussions and the resulting report, the committee believes that the material presented here will help identify specific areas for more in-depth inquiry by all the stakeholders in STM journal publishing and in scientific communication more generally.

Edward Shortliffe
Committee Chair
Paul Uhlir
Project Director

Acknowledgments

The Committee on Electronic Scientific, Technical, and Medical Journal Publishing and Its Implications would like to thank the following individuals (in alphabetical order) who made presentations during the symposium (see Appendix A for the final symposium agenda): Hal Abelson, Massachusetts Institute of Technology (MIT); Bruce Alberts, National Academy of Sciences; Kent Anderson, *New England Journal of Medicine;* Malcolm Beasley, Stanford University; Robert Bovenschulte, American Chemical Society; Monica Bradford, *Science;* Patrick Brown, Stanford University; Brian Crawford, John Wiley & Sons; James Duderstadt, University of Michigan; Joseph Esposito, SRI Consulting; Michael Jensen, Harvard Business School; Michael Keller, HighWire Press; David Lipman, National Center for Biotechnology Information; Wendy Lougee, University of Minnesota; Richard Luce, Los Alamos National Laboratory; James O'Donnell, Georgetown University; Paul Resnick, University of Michigan; Bernard Rous, Association for Computing Machinery; Alex Szalay, Johns Hopkins University; Gordon Tibbitts, Blackwell Publishing USA; and Ann Wolpert, MIT.

The committee also would like to express its gratitude to the guidance group for this project, which was formed under the Committee on Science, Engineering, and Public Policy. Members of that group included, James Cook, Washington State University; Paul Torgerson, Virginia Polytechnic Institute and State University (retired); and Edward Shortliffe, Columbia University Medical Center, Columbia University.

This volume has been reviewed in draft form by individuals chosen for their diverse perspectives and technical expertise, in accordance with procedures approved by the NRC's Report Review Committee. The purpose of this independent review is to provide candid and critical comments that will assist the institution in making its published report as sound as possible and to ensure that the report meets institutional standards for objectivity and evidence. The review comments and draft manuscript remain confidential to protect the integrity of the deliberative process.

We wish to thank the following individuals for their review of this report: Martin Blume, American Psychological Society; Karen Hunter, Elsevier Health Services; Justin Hughes, Cardozo Law School; James Neal, Columbia University; Andrew Odylzko, University of Minnesota; and Carol Tenopir, University of Tennessee at Knoxville.

Although the reviewers listed above have provided constructive comments and suggestions, they were not asked to endorse the content of the report, nor did they see the final draft before its release. The review of this report was carried out in accordance with institutional procedures and all review comments were carefully considered. Responsibility for the final content of this report rests entirely with the authoring committee and the institution.

Finally, the committee would like to recognize the contributions of the following National Research Council staff. Paul Uhlir, director of the Office of International Scientific and Technical Information Programs, was the project director for the symposium and principal editor of the committee's report; Julie Esanu, program officer for the Office of International Scientific and Technical Information Programs, helped organize the symposium and edit the report; Alan Inouye, interim director of the Computer Science and Telecommunications Board, and Robin Schoen, program officer for the Board on Life Sciences, provided advice on the project; Kevin Rowan, project associate for the Committee on Science, Engineering, and Public Policy, provided project support for the May symposium; and Amy Franklin, senior program assistant for the Board on International Scientific Organizations, assisted with the production of this report.

Contents

Executive Summary   1

1. Introduction   7

2. Costs of Publication   9

3. Publication Business Models and Revenue   20

4. Legal Issues in the Production, Dissemination, and Use of the Journal Literature   40

5. What Is Publishing in the Future?   48

6. What Constitutes a Publication in the Digital Environment?   56

7. Symposium Wrap-Up   67

Appendixes

A Symposium Agenda   75

B Biographical Information for Speakers and Steering Committee Members   80

C Symposium Participants   94

Executive Summary

The Symposium on Electronic Scientific, Technical, and Medical (STM) Journals and Its Implications addressed these issues in five key areas. The first two—costs of publication and publication business models and revenue—focused on the STM publishing enterprise as it exists today and, in particular, how it has evolved since the advent of electronic publishing. This was followed by a review of copyright and licensing issues of concern to the authors and to universities. The final two sessions looked toward the future, specifically, at what publishing may be in the future and what constitutes a publication in the digital environment.

COSTS OF PUBLICATION

The initial session identified the main elements of expenses (both print and electronic) for many STM journal publications. It was quite clear from the presentations that understanding all the cost elements in the budgets of different publishers is not straightforward and that a comparative analysis, in particular, would be very difficult to do well. At the same time, some speakers indicated that such a study, if done carefully and authoritatively, could be very useful in adding some rigor to the ongoing debate about the high cost of journal subscriptions and the value that publishers bring to the process of scientific communication.

Other cost-related issues that were discussed included the creation and operation of digital archives and the digital conversion of back sets; the costs of new technology and related cost-containment strategies and the difficulty of moving from print to electronic-only versions; strengths and weaknesses of the peer-review process; cost issues specific to small and mid-sized societies; and the vulnerability of secondary and tertiary publishers.

PUBLICATION BUSINESS MODELS AND REVENUE

This, understandably, was the most contentious part of the symposium. A number of trends in commercial STM journal publishing were described at the outset. These include the bundling of publications by major publishers in "big deals"; the consolidation of publishers and the targeting of downstream competitors (secondary publishers and subscription agencies) and of vulnerable competitors; diversification of the customer base to more business clients (and a concomitant emphasis on applied research and engineering journals); and market responses to open-access trends, including the creation of meta-content (e.g., documentation and search engines for the open-content resources) and a shift to Web services (e.g., substitutes for the publication of fixed content in print by providing online software, processing, and services for users).

This discussion of commercial publishing trends was followed by the perspectives of a university librarian, a commercial journal publisher from John Wiley & Sons, Inc., and an open-access representative from the Public Library of Science (PLoS). The library overview included some statistics and anecdotal information about the responses of libraries to rising subscription costs and the bundling efforts of commercial publishers; the implications for libraries of changes in electronic journal formats and content, and of digital scientific communications more broadly; and the changing role and influence of libraries in the digital publication context.

The description of the commercial subscription-based model used at John Wiley & Sons highlighted the benefits to readers from this traditional approach and the reasons why Wiley would not switch to the author-pays, open-access model.[1] This was followed by the description of the PLoS

---

[1]According to the definition presented by the Public Library of Science later in this report, an open-access publication is one that meets two conditions. The first is that the copyright holder (either the author or the publisher, if the copyright has been transferred to the publisher) grants to the public a free, irrevocable, perpetual right of access to, and a license to copy, distribute, perform, and display the work, and to make and distribute deriva-

model, including the rationale for reconceptualizing the STM publishing business model on the Internet, the definition of "open access" used by the PLoS, and the advantages of this approach for science.

A vigorous and informative discussion ensued among the panelists and with the audience about the relative merits of the traditional user-pays publishing model versus the author-pays PLoS model. Other related issues were discussed, including the practical difficulties in transitioning to the open-access publishing model, support of the open-access publishing approach by research sponsors, the effect of different publishing business models on the long-term preservation of digital journals, and advertising revenues in electronic publishing generally.

The results of these discussions appeared to be inconclusive. On the one hand, commercial (and professional society) publishers clearly add considerable value to the process of formal scientific communication, and the viability of the author-pays, PLoS type of open-access model is still untested and its future success uncertain. On the other hand, the restricted, subscription-based model clearly has great inherent social costs in comparison with the immediate, free access by any and all users of the information worldwide that the open-access publishing model makes possible. Many participants believed that we are in a period of important experimentation, in which the open-access approach will be tested and refined and in which the traditional publishers will try new approaches and attempt to add more value to STM products and services. A greater differentiation between the practices of commercial and society publishers also may be expected. For example, there are hybrid approaches being implemented in the traditional subscription-based publishing community, mostly by the not-for-profit STM publishers. Moreover, there are other open-access approaches such as self-archiving by authors and open institutional repositories, as discussed in other sessions of the symposium.

---

tive works in any medium for any purpose. The second condition is providing readers with open access to the work. Authors or publishers achieve open access by making a complete version of the article and all supplemental materials available in some suitable standard electronic format, deposited immediately upon publication in at least one internationally recognized, independent online repository that is committed to open access.

LEGAL ISSUES IN PRODUCTION, DISSEMINATION, AND USE

The focus in this session was on copyright and on licensing issues in the traditional publishing business model. With regard to copyright, there are divergent practices at universities as to whether the university or the author owns the copyright to publications, and in the various derogations from those rights. The question of transfer of copyrights from the author to the publisher, and the limited rights granted back by the publisher to the author, was discussed as well.

The licensing issues pertain to the terms and conditions that publishers and libraries negotiate for site licenses, and to licenses between authors and publishers. The libraries have continued to experience two significant problems in negotiating site licenses, both related to the overall problem of access restrictions: the various limitations and prohibitions on "interlibrary loans" of electronic copies and the problem of long-term preservation of material that is electronic-only under the restrictive licensing regime. Both of these factors inhibit libraries from switching to electronic-only subscriptions. As to the licenses between authors and publishers, two models were suggested that serve the author's interests better: either retain copyright, while licensing the publisher to deploy the work in all ways that the publisher needs for effective publication and dissemination, or transfer copyright to the publisher, with more rights reserved to the author, such as permission to redistribute the work.

Issues raised in the general discussion included a description of additional problems with the transfer of copyrights by authors in universities (i.e., the author may not own the copyright under the university policy but may not know that, thereby signing void copyright transfer agreements) and significant problems associated with university work-for-hire approaches to academic publications. The burdens for small publishers in developing countries from licensing practices and from restricting access were also discussed.

WHAT IS PUBLISHING IN THE FUTURE?

The final two sessions of the symposium looked more toward the future, by identifying some of the technology-enabled trends, processes, and projects that are indicative of what may be possible and what may perhaps become more widely adopted. As was noted in the introductory comments for this session, it is quite clear that the digital revolution is changing the traditional processes of many knowledge-intensive activities, in particular STM publishing and scholarly communication more generally. The various functions—whether metadata creation, credentialing review, or long-term stewardship—can be separated or disaggregated, and players different from those who traditionally have carried out these tasks can, in theory, perform them. Publications can now exist in many intermediate forms, and we are moving toward more of a continuous-flow model, rather than a discrete-batch model. The raw ingredients—the data, the computational models, the outputs of instruments, the records of deliberation—can be online and accessible by others and can be used to validate or reproduce results at a deeper level than traditionally has been possible. Third parties—particularly in an open-access, open-archives context—can then add value by harvesting, mining, enriching, and linking selected content from such collections.

The presentations in this session of the symposium identified some of the social processes, specific pilot projects, and the challenges and opportunities that may provide the basis for future "publishing processes," which ultimately may be more holistically integrated into the "knowledge creation process." For example, there are emerging open recommender and reputation systems that use the online environment to get broader public feedback and to develop new indicators of user behavior. Although there are potential problems, such as "gaming" the system, eliciting early evaluations, and "herding," whereby later evaluators are unduly influenced by previous evaluators, there are some experiments that could be tried in the STM publishing context. Preprint servers, such as the well-known e-Print arXiv established initially for the high-energy physics community by Paul Ginsparg at the Los Alamos National Laboratory, have now been adopted by some other fields. Open institutional repositories, such as the MIT OpenCourseWare project and the MIT-led DSpace consortium for the deposit of various types of research materials in an openly accessible archive, provide additional examples of innovative projects and models.

Despite the exciting possibilities raised by these different initiatives, the subsequent discussion raised a number of potential difficulties in implementing some of these new approaches. Some of the issues raised included caution about an over-reliance on statistical indicators or metrics in judging the quality of information or of publishing activities; the relative merits of the traditional, confidential peer-review process and a more open review system; the constraints of the discrete-batch, print model on adopting some of the more open, continuous-flow processes; and the inherent tensions between the creation of various open archives and the traditional publishers.

WHAT CONSTITUTES A PUBLICATION IN THE DIGITAL ENVIRONMENT?

The final session built on the technology-oriented concepts and processes introduced in the prior session. Publication used to refer to the act of preparing and issuing a document for public distribution. It could also refer to the act of bringing a document to the public's attention. Now, publication means much more. It can refer to a document that is Web-enriched, with links, search capabilities, and potentially other services nested in it. A publication now generates usage data and provides many other functions. This session examined three innovative examples—the Signal Transduction Knowledge Environment of *Science*, the publishing of very large data sets in astronomy on Web sites and through the International Virtual Observatory initiative, and genomic data curation at the National Center for Biotechnology Information and the integration of those data with the scientific literature.

Issues raised in the discussion included the need for federal coordination and investment in the cyberinfrastructure to maximize the opportunities for information integration and knowledge creation, some of which were identified in the final two sessions; the difficulties of quality control and review of data in very large or complex databases, particularly in the biological sciences; restrictions on data mining in proprietary STM information that is based on publicly funded research; the opportunities for knowledge discovery from the open publication of large and complex data sets; the transformation of the archiving function in the knowledge discovery process; lost opportunities associated with insufficient people and resources focused on the avalanche of data in all disciplines; and the positive role of the journal publishers in the successful development of databases in molecular biology.

1

Introduction

The main focus of this symposium was how different business and distribution models for scientific, technical, and medical (STM) information publishing are changing in the face of digital technology developments. How do the emerging models address the need of the scientific communities for the widest possible long-term access to such information? In developing the symposium's program, the planning committee was mindful of the broad, ongoing changes in scientific research, funding, and goals—changes that stimulate, and are stimulated by, new forms of scholarly communication. The challenge was to identify issues and problems that the STM communities need to control and resolve in order to exploit the remarkable and growing opportunities offered by the rapidly evolving computer and networking technologies.

Indeed, the current situation can perhaps be described as a "chaos of concerns."[1] The advances in digital technology are producing radical shifts in our ability to reproduce, distribute, control, and publish information. Yet, as these advances increasingly become an integral part of scientific activity, they tend to conflict with some of the existing practices, policies, and laws that govern traditional publishing.

---

[1]Portions of this Introduction are based on the keynote presentation by James Duderstadt, president emeritus and University Professor of Science and Engineering Millennium Project, University of Michigan.

The issues are complex, in part, because the stakeholders are so many, so varied, and with different agendas. Those who fund research want to see that the information is advanced and made available to the public. The authors, reviewers, and part-time editors do not charge the publishers for their labor. They are motivated to contribute to the public good and to receive public recognition for their ideas, but of course they also have other rewards, not the least of which is tenure. The journal publishers, as intermediaries, while they do not pay for the material that they publish, do add significant value, and provide the scholarly output in a useable, published form. Libraries, similarly, provide an intermediary function by paying the increasingly expensive subscription fees, though they typically do not charge for subsequently providing access to the journals. And, of course, the end users—who are also primarily the originators of the STM journal literature—either pay for their own personal subscriptions or obtain the resources free through libraries.

All these stakeholders have their own needs and objectives in the evolving process of scholarly communication, many of which are congruent, but some that now conflict, and all of which are being continually redefined by technological, institutional, financial, and other changes. Current uncertainties in the responsibilities for digital preservation and archiving functions are but one example of this. The symposium therefore provided a forum for identifying and discussing some of the key pressure points and areas of disagreement by bringing together expert representatives of these major stakeholder groups.

The symposium began with speakers in the first two sessions examining cost and revenue aspects of different business models in publishing from the perspectives of some of the main types of stakeholders in the STM publishing process. The third session provided a brief overview of legal issues in the production, dissemination, and use of STM journals, focusing on copyright and licensing by authors, universities, and publishers. Speakers in the next two sessions then looked toward the future, speculating about what publishing might be in the future and what constitutes publication in the digital environment. The final session provided the perspectives of several invited speakers on the symposium discussion. This summary highlights many of the key issues identified during the course of the proceedings, in both the invited presentations and in the subsequent discussions with the expert audience.

2

Costs of Publication

Electronic publishing has not only revolutionized the publishing industry, but it has also tremendously changed the fundamental economics of the STM journal business. Many new issues affect both for-profit and not-for-profit publishers. For publishers, the costs of journal production appear to be increasing rapidly, much more than if publishers had stayed with print alone. Two obvious factors are responsible for these rising costs: the cost of new technology and the increasing volume of publishing. Thus, it is natural to begin with a consideration of the costs of STM publishing.

MAIN ELEMENTS OF EXPENSE BUDGETS FOR SOME STM JOURNAL PUBLICATIONS[1]

Many STM journal publications have the same or similar cost components. First are the processing costs for the contents (including articles,

[1]This chapter is based on the remarks of Michael Keller, Ida M. Green University Librarian at Stanford University and publisher of HighWire Press, who provided a keynote "Overview of the Costs of Publication," as well as on the comments of invited speakers— Kent Anderson, publishing director at the *New England Journal of Medicine*; Robert Bovenschulte, director, Publications Division, American Chemical Society; Bernard Rous, deputy director and electronic publisher, Association for Computing Machinery; and Gordon Tibbits, president, Blackwell Publishing USA. Unlike the summaries of the subsequent sessions, the comments made by several participants in the audience are incorporated into the presentation material as well.

reports of research results, and methods of scholarly investigation) of STM journals, in several subcategories:[2] (1) manuscript submission, tracking, and refereeing operations; (2) editing and proofing the contents; (3) composition of pages; and (4) processing special graphics and color images. Internet publishing and its capacity to easily deliver more images, more color, and more moving or operating graphics have made this category of expense grow rapidly in the past decade. The second category of expense is not only a familiar one, but is also one of two targets for elimination from publishers' costs: This includes costs for paper, printing, and binding, as well as mailing. A third category is the cost of Internet publishing services. These are new costs and include many activities performed mainly by machines typically maintained by highly paid technical support staff, though in some situations the publishing staff performs quality control pre- and post-publication to check and fix errors that may have been introduced into the publishing process. The elements of all the above costs vary tremendously among publishers and Internet publishing services. The fourth cost category—publishing support—includes items such as supporting facilities, marketing, and other miscellaneous expenses. The final category is the cost of reserves.

The results of a recent sampling by Michael Keller of the changes in six not-for-profit publishers' costs over the past decade indicated several trends. It appears the publishers now exercise much tighter controls over their budgets than they did 10 years ago. Although the rate of change in editorial costs has not changed much (e.g., costs continue to scale with the number of submissions and inflation in salary and benefit costs), printing, paper, and binding costs are down, at least on a unit basis. The new costs associated with electronic versions of the same publications also can be significant.

In short, there is a dynamic balancing act with regard to publishers' costs in the Internet era, with some costs increasing and some decreasing. Most intriguing, however, is the possibility of removing substantial amounts of publishing costs by switching to electronic-only journals delivered over computer networks, thereby eliminating printing, binding, and mailing paper copy to all subscribers.

---

[2]The cost budgets for *Science* and *Nature*, for instance, would have more elements than specified. And costs for secondary and tertiary publications include different elements than these. These sorts of costs in STM publishing are not covered here.

DIGITAL ARCHIVES

Successful creation and operation of true digital archives—protected repositories for the contents of journals—would permit the removal of the printing, binding, and mailing costs. True digital archives will have their standards and operational performances publicly known and monitored by publishers, researchers, and librarians alike. The operations and content of such archives will need to be audited regularly. The annual costs are not yet well understood; they can range widely, from only tens of thousands of dollars per year for a simple approach such as LOCKSS (Lots of Copies Keep Stuff Safe), which are not formal archives,[3] to $1.5 million per year for a large archive at a major university. Publishers, especially not-for-profit publishers, feel a very strong obligation to preserve that digital heritage, although there is no consensus about exactly who should do so: they themselves, libraries, some other third-party archivers, or some combination of all of them.

CONVERSION OF BACK SETS

Another cost currently confronting many publishers is the conversion of back sets of print journals to digital form, providing some level of metadata and word indexing to the contents of each article, and posting and providing access to the back sets. HighWire Press has done a review of the costs of converting the back sets of its journals. They estimate that about 20 million pages could be converted and that the costs of scanning and converting pages to PDF, keying headers, loading data to the HighWire servers, keying references, and linking references could approach $50 million, or about $150,000 per title. Most of that sum is devoted to digitizing companies and other sub-contractors of HighWire Press.

If all this retrospective conversion of back sets occurred in one year, HighWire would have to spend internally about $250,000 in capital costs and about $300,000 in initial staff costs, declining to annual staff expenditure of perhaps $250,000 or $275,000 thereafter. On average, for the 120 publishers paying for services from HighWire that would mean about an additional $2,500 in new operating costs to HighWire Press each year. In other words, the increase in annual costs to publishers for hosting and pro-

---

[3]See http://lockss.stanford.edu.

viding access to the converted back sets would be a fraction of 1 percent of their current expenses each year. These figures, of course, omit any costs for digitizing and other services provided by contractors and subcontractors.

Although the costs of back-set conversion are high, the experience of HighWire suggests that the payoff could be 5-10 times more use of articles in the back sets than is presently experienced. Articles on the HighWire servers are read at the following rates: Within the first 3 months of issue, about 95 percent of all articles get hits (this presumes that a hit means that somebody is actually reading something). In the next 3 months, that is, when the articles are 4-6 months old, slightly less than 50 percent of all articles get hits. And when articles are 10 months or more old, an average of only 7-10 percent of all articles get hits. However, that rate of hits seems to persist no matter how old the online articles are.

Based on citation analyses, only 10 percent of articles in print back sets older than the online set of digital versions get cited, though not necessarily read. That they should do so is entirely consistent with the belief commonly held since 2001 by publishers associated with HighWire that the version of record of their journals is the online version. This is leading many publishers to digitize the entire run of their titles as the logical next step. In any case, unless other sources of funds are forthcoming, the costs of back-set conversion will become a temporary cost in the expense budgets. Other STM journal publishers, however, have indicated that their back-set conversion and subsequent maintenance costs have been considerably higher.

THE VALUE OF MAINTAINING THE EXISTING STM JOURNAL PUBLISHING INFRASTRUCTURE

Some stakeholders in the STM journals chain believe that in the future articles will be delivered freely to all users, through a diffuse distribution scheme based on authors simply posting their articles on their own Web sites or on an archive like the Cornell e-print arXiv. Google, or more specialized search engines, would be used by readers to access relevant articles on demand. This approach, they assert, could obviate the need for today's expensive publishing apparatus.

Nonetheless, most communities in STM assert that there remains a strong requirement for formal peer review of STM journal content, and it seems clear that many of the functions provided by good publishers are highly valued and demanded by scientists. There is no question that providing relevant, reliable, and consistent levels of content in journals costs money. Highly distributed, diffuse STM publishing with sketchy peer review, dependent upon new search engines to replace the well-articulated scheme of thematic journals and citations in a multidimensional web of related articles, represents a descent into information chaos.

BALANCING THE NEED FOR SPEED WITH THE MAINTENANCE OF QUALITY

The pressures to distribute information rapidly are growing, even as high-quality content must be maintained. For example, the *New England Journal of Medicine* (NEJM) publishes information about health, and if it publishes erroneous articles, these can have serious impacts. Recently, the journal published a set of articles about SARS, Sudden Acute Respiratory Syndrome, within two weeks of receipt from the author, and these were completely peer reviewed, edited, and illustrated papers. They were translated into Chinese within two days of initial publication and distributed in China in the thousands in print, with the hope that they would make a major difference.

COSTS OF NEW TECHNOLOGY

With the expansion of technology costs, publishers are delivering a much more valuable product to their users. Enormous new functionalities are being made available to scientists. The access to information is swift and convenient, improving scientists' productivity. Technology now imbues all facets of publishing—from author creation and submission, all the way through to peer review, production and editing, and output and usage. The costs of the technology are not just related to the Web, but apply to all the other technical systems that publishers must create and integrate.

The various technological enhancements can have value, and some publishers have been surprised at how much demand there has been for new applications. There is thus pressure on the publishers to compete with other publishers' innovations. On the one hand, many of the recent innovations that drive up publishers' IT costs are very little used and not of great value. On the other hand, it is very hard to predict utility in advance of introducing such innovations. Some technologies that are developed on the margin eventually become quite popular. Moreover, even an innovation that is not much used today may turn out to be one that is very valuable 5 or 10 years from now. So, publishers are experimenting and attempting to stay abreast of technological developments to remain competitive.

The experience of HighWire Press, for example, has been that those publishers who first define and create advanced features subsidize some of the costs for publishers that follow. At the same time, those early adopters reap the benefits of innovation in attracting authors and readers. Eventually, many of the innovative features become generally adopted, and usually at a lower cost of adoption than paid by the innovators to innovate.

THE DIFFICULTY OF MOVING FROM PRINT TO ELECTRONIC-ONLY VERSIONS

There is not only a continuing demand for print versions from some customers, libraries, and users, but also from the publishers themselves, because the profit margins that are realized from the print side are actually necessary for some of them at this time. For a variety of reasons, in some communities there are many readers for whom online reading and searching are not presently good options. Because of the preservation issue, most librarians are not yet ready to give up print either.

Yet for many research communities, especially in the basic sciences that develop rapidly, there is real promise for dropping the print versions of journals altogether. That does not necessarily mean, however, the total discontinuation of those paper, printing, and binding costs. Once print versions are eliminated, there might be a cost savings for publishers of 15, 20, or 25 percent. Ending print versions of journals is probably a worthwhile goal. The concern, however, is that the costs of managing the rising volume of publication will wipe out whatever transitory gains there may be from saving on print costs.

STRENGTHS AND WEAKNESSES OF THE PEER-REVIEW PROCESS

The peer-review process is both a significant cost and the highest value added by the journal publishers. Although many STM journals have rigorous peer-review procedures, the peer-review process at the *New England Journal of Medicine* (NEJM) was provided as just one example. First, internal editors at the NEJM review incoming article submissions, judging them for interest, novelty, and completeness. Papers that are of initial interest are sent to two to six external peer reviewers. Returned reviews are used to judge whether the submission will move forward. If it does, it is brought before a panel of associate editors, deputy editors, and senior editors; the paper is explained, questions are asked, and the work is judged by that group, usually during a very thorough discussion. Having passed the editorial board, the article is sent for both a statistical and a technical review. Queries are brought to the author, who must respond. The peer-review process lasts anywhere from a few weeks to a few years. Sometimes the NEJM asks the authors to either complete experiments or to provide additional data.

As far as weaknesses in the peer-review process, or ways in which it could be improved, one of today's concerns is that time pressures in medicine are so great that finding willing peer reviewers is increasingly difficult. The situation could be improved by educating the scientific community to understand the value of this interaction in the STM publishing process, reflecting that value in the academic rewards system.

As the number of submissions has risen, the number of people available to provide dependable reviews of articles has not increased. Publishers are calling upon the same people time and time again, and they perform peer review for the most part without compensation.

COST-CONTAINMENT STRATEGIES

If the only imaginable strategy for supporting the costs of publication is to increase subscriptions paid for by the libraries and laboratories, then frankly, there is no future for STM publishing. If the publishers are to survive, they must have strategies for reducing costs or increasing income from other sources of revenue.

Even as the costs of journal subscriptions have increased, and publishers' costs have increased, the cost per person of accessing the body of research articles has plummeted dramatically in the electronic context, and this trend is likely to continue. There are many potential ways of collaborating on technologies that can bring production costs down dramatically. For example, using open-source software can reduce development costs significantly. There is no reason why publishers should develop unique, proprietary online systems.

Moreover, with innovations from commercial vendors such as Adobe and Microsoft and simple services such as e-typesetting in China, India, and other places, the costs for both small and large publishers alike can be lowered. The use of simpler, standard formats that all publishers agree to adhere to and avoiding the use of very complex Web sites are also effective cost-saving measures.

VULNERABILITY OF SECONDARY AND TERTIARY PUBLISHERS

In the next decade, the segment of STM publishing that appears to be most at risk is the secondary publishing industry, that is, the abstracting and indexing services and the tertiary publishers, those producing review articles long after the leading-edge researchers have studied the most useful articles.

Secondary publishers are particularly vulnerable, because they are being overtaken by numerous sophisticated search engines and by development of peer-to-peer data about usage patterns—the sort of features and functions provided by Amazon.com, for example. The secondary publishers, therefore, must find ways to become more effective, more precise, as well as more general. They need to seek out different approaches, particularly automated services. However, if STM information does move toward a highly distributed and diffuse dissemination mode, in which authors place their contributions on individual servers, there could be a huge role for secondary publishing.

The secondary publishers are vulnerable now, not necessarily because secondary publishing is vulnerable, but because they are competing with primary journal publishers who are, perhaps inadvertently, providing secondary services. Secondary publishing is in some sense merging together with primary publishing, especially in the aggregator business.

ELECTRONIC PUBLISHING BY SMALL AND MID-SIZED SOCIETIES

For a small or mid-sized society it may be very risky to stop delivering printed journals, which are perceived by society members as a strong benefit. One reason that some smaller societies can manage to publish electronically involves the compounding effect of information exchange that occurs at the meetings that they have with their electronic service providers. Technologically, many scientific societies cannot afford to publish electronically on their own.

Societies also have varying business circumstances. For example, the Endocrine Society is tied to its print clinical journal because it generates $2 million per year of pharmaceutical advertising. Moreover, not all of the smaller societies are yet able to convert to electronic publishing. The American Psychological Society several times tested in quite elaborate detail whether it could publish its own journals online, and recently has decided it cannot. In this case, the risks, particularly for a small organization with modest reserves, simply cannot be taken.

The suggestion that there be differential service levels poses difficulties as well. The Association for Computing Machinery (ACM) moved to multiple service levels in order to maintain a benefit for personal membership in the organization, as its institutional subscriptions. But differentiating service levels adds real cost and complexity to any journal publishing system. It also adds complexity for the end user, and additional communications activities for the society that needs to explain its tiers to the different users.

SOME UNANTICIPATED COSTS OF ELECTRONIC PUBLISHING

Customer support costs for electronic publishing have been substantially different from the print paradigm. Not only is there a larger volume and variety of customer complaints and requests for improvements, but also the level of knowledge and the expertise required to answer user questions call for much more costly staff. The cost of sales currently is higher. The product is different and the market is shifting. Many publishers, such as the ACM, no longer sell subscriptions; rather, they license access. Additional personal contact is required to market and sell an electronic site license, so this requires additional and expensive sales personnel. Also, digital services are built on top of good metadata, and metadata creation and development costs are high. The richer the metadata, the higher the costs. Subject classification can be costly as well. Finally, what were thought to be some upfront, one-time costs for electronic publishing have turned out to be recurring costs, and some recur with alarming frequency.

REASONS WHY THE COSTS OF ELECTRONIC PUBLISHING ARE POORLY UNDERSTOOD

The costs of electronic publishing are not at all well understood, and there are some very good reasons why this is the case. First, as has already been mentioned, electronic publishing is not a single activity.

Second, electronic publishing costs remain fuzzy because we still live in a bimodal publishing world. Some direct expenses can be charged to either print or electronic cost centers. Where a publisher allocates costs often depends on the conceptual model of a particular publishing enterprise.

Third, the decisions about whether to charge costs to a print or to an electronic publication can also be part of a political or business process. There are times when it is desirable for the publisher to isolate and protect an existing and stable print business, and in that circumstance the publisher will attribute any new costs to the digital side. Or, the publisher may desire to minimize positive margins on the digital side, in order to avoid debate over the pricing of electronic products. At other times, the need to show that the online publication has taken wings, is self-sustaining, and has a robust future can tilt the allocation of all debatable charges to the print side.

Fourth, it is difficult to compare print and electronic costs because the products themselves are just not the same.

Fifth, accounting systems sometimes evolve more slowly than shifts in the publishing process. New costs appropriate to online publications are sometimes allocated to pre-existing print line items.

Sixth, electronic publishing has not yet reached a steady state. There is a great deal of development, some of which lowers costs and some of which increases them.

These are some of the reasons why the costs of electronic publishing remain obscure, and also why a study of such costs would be both difficult to carry out and very important to attempt.

WHAT WOULD A STUDY OF JOURNAL COSTS ACCOMPLISH?

It is critical for the research community to have a common understanding of the problem set. A high-quality study of costs and benefits of electronic journal publishing from the birth of the World Wide Web to the present could elevate the level of discourse among the stakeholders. Such a study could document, in a neutral way, the profound transformation of an important aspect of the national research effort. That there is likely to be as much change in the next 10 years as in the past decade does not obviate the need for the study. Furthermore, such a study may help to develop new strategies or evolve current ones for accommodating needs of scientists and scholars in reporting their research findings and for ensuring the long-term survival of the history of science, medicine, and technology. How experimental business models might provide competitive pressure on traditional business models and pricing is a topic for discussion and examination over time.

If publishing costs are to be studied well, however, there must be an acknowledgment of the diversity of the publishing landscape. If only a few publishers participate, the selection bias could drive the study to the wrong answers. It also is necessary to consider which components need to be analyzed. The questions to be asked must be properly framed and a reasonable control group selected. In short, a study of this nature could be valuable, but it needs to be well designed, rigorously conducted, and carefully interpreted.

3

Publication Business Models and Revenue

Scholarly publishing in any medium requires substantial resources beyond content creation costs. Publishers are providers of value-added products and services. They perform crucial functions by producing documents in different media, performing editorial and design work, marketing the material, and connecting readers to writers, and so forth. All of those functions involve costs. Even a not-for-profit publisher has to at least recover costs and generate a reserve.

It is useful, therefore, to examine sources and types of revenue, ways of raising revenue, and different business models, particularly in a world where digital publishing is becoming much more the norm. The business models are related to what information is being published, for what audience, and how it will be accessed. In a digital world we no longer need to have a single standard mode (i.e., the journal). We can think about presenting information in lots of different ways and repackaging it and distributing it in different combinations.

What is the role for government in this process? Much of what we are talking about—scientific, technical, and medical information and scholarly communication includes information that benefits the general public either directly or indirectly, far beyond the community of scientists and scholars who are using it. There is a public interest in the dissemination of knowledge, in addition to its creation.

Of course, the overriding question for this symposium is what impact the digital publishing world is going to have on science itself, that is, on the scientific enterprise. For instance, how are different business and access models going to affect the quality and productivity of science, collaboration at a distance, access for developing countries, the professional review and career process, peer review, and other aspects of scientific research?

The discussion that follows looks at the major business models from several stakeholder perspectives. It begins with an overview of trends in the commercial STM journal publishing industry, followed by a perspective from the library community, which serves as the intermediary between STM publishers and the academic user community. Two contrasting publishing paradigms are presented next: the traditional "reader-pays" model, as implemented by John Wiley & Sons, Inc. and the newer "open-access," "author-pays" model of the Public Library of Science. The section concludes with a review of issues raised in the discussion with the expert audience.

TRENDS IN COMMERCIAL (FOR-PROFIT) STM JOURNAL PUBLISHING[1]

Consolidation of Publishers

The for-profit STM publishing market is "mature," which means that there is only modest growth in revenue, and it is difficult for new players to enter that market. In a time of immense change, publishers are experimenting with new and different ways of working in this marketplace. We should expect, for example, to see consolidation among publishers continue, even as the customers (mostly libraries) continue to launch antitrust actions against buyouts or mergers between significant STM players. That said, it is highly unlikely that any scientific discipline will have more than two or three information providers in the years ahead.

Moreover, the form that consolidation will take may change. In addition to the already familiar phenomenon of big STM companies buying up smaller ones, we should look for smaller companies to link together in an attempt to provide levels of service and functionality similar to those offered by the big companies.

---

[1]The information in this section is based on the remarks of Joseph Esposito, president and chief executive officer of SRI Consulting.

Bundling

A mature market should be expected to intensify downward pressure on journal prices. This is already true in academic research institutions, where the open-access movement is being created in part to put pricing pressures on STM publishers, both for-profit and not-for-profit. Publishers will likely counter such pressures by targeting the market share of other publishers, rather than looking for significant increases in library budgets. For example, Reed Elsevier is now offering libraries access to previously unsubscribed journals, not by charging for each journal separately, but simply by insisting on an increase in total expenditures over the prior year. This practice has been referred to as supersizing, or the big deal. In other contexts, it is called bundling or tying. Bundling will have the effect of greatly increasing the number of Reed publications available through particular libraries, at the expense of having less well positioned publishers lose those customers entirely.

Downstream Value Migration

We also should expect commercial publishers to seek so-called downstream value migration and to target competitors that for various reasons are thought to be vulnerable. These "competitors" may include, for example, former partners such as secondary publishers or subscription agencies. By moving downstream, more publishers will attempt to disintermediate[2] the wholesalers and reap the wholesalers' marginal revenue. Disintermediation strategies that do not provide significant new value to end users are probably ineffective, but that does not mean publishers will not try such strategies.

Targeting Vulnerable Competitors

For-profit publishers are likely to target the not-for-profits more aggressively in the future. The reason is that the not-for-profits may be per-

---

[2]"Disintermediation" is the process by which new Internet-based products and services replace products or services that existed in the pre-Internet era, particularly ones that serve as intermediaries between the provider of a product or service and the end user.

ceived to be slower to respond to technological capabilities and to be less competitive, though they often have substantial goodwill in the marketplace. There are many prospective targets among the small university presses and learned societies.

Creation of Meta-Content

Metadata are often defined as "content about content," and they can be exploited or created as bibliographies, indexes, and, most important of all, through search engines. Publishers are keenly aware of open-access publications and are looking for ways to make money from them. Open-access publications are, by definition, available for any individual organization to use without permission or fee. Thus, one way for publishers to use open access is to create search engines for open-access content. Even more powerful is to integrate open-access content with proprietary content for search purposes. In other words, open-access publications provide publishers with lower costs for content development while enabling for-fee services. From an economic point of view, copyright transfer to publishers is unnecessary for supporting publishing profits.

The Shift to Web Services

The most significant economic response to open access is likely to be in the creation of Web services, in the form of dynamic substitutes for the publication of fixed content in hard copy. In a Web service, a publisher will provide online software that manipulates or processes data that are uploaded to it by a user. The user creates the content and then pays the service provider for the online processing. Copyright is irrelevant for models like this, even as the economic potential is very great.

Diversification of Customer Base

If the academic channel is mature (i.e., lacks the potential to grow rapidly, if at all) publishers will seek new sales channels. The most likely one, because of its size and creditworthiness, is sales to businesses such as engineering, chemical, or pharmaceutical companies. Thus, publishers' capital investment may shift from pure research publications toward applied research and engineering.

In 5-10 years, open-access publications will coexist with proprietary ones, and we will witness ingenious publishing strategies designed to extract economic gain even in the absence of a proprietary distribution model.

THE LIBRARY PERSPECTIVE[3]

The scholarly journal has existed for more than three centuries. The journal provides a trusted place to document discoveries, disseminate ideas, and codify prestige. This three-century tradition will not easily change.

Responses of Libraries to Recent Cost and Marketing Trends

Research libraries are the intermediaries between two types of economies. They buy content in a market economy and make it available in the nonprofit, academic sector. Thus, libraries are often caught in the clash between the market and the gift economies. In this position, the libraries have witnessed decades of journal price increases, with average annual increases for the past 5 years being around 8 or 9 percent. It is a very inelastic market, because as the prices increase, libraries are not able easily to withdraw or cancel the costliest journal subscriptions. Data over the past 15 years show that journal prices have increased by 215 percent, yet libraries canceled only about 5.1 percent of their subscriptions. Despite the apparent lack of elasticity in this market, the ability of libraries to continue to afford all research content, in the face of escalating STM journal prices, is certainly cause for concern.

STM journal price increases and inelasticity have increased in the past 2 years as a result of two developments. The first is that the publisher strategy of the so-called big deal—the multiyear, all-titles packages sold by many publishers to libraries and consortia—has begun to unravel. Some research libraries intend to withdraw from journal package arrangements, because of budget reductions and the low or non-use of a significant proportion of titles in the package. Libraries are beginning to push for more finely tuned licensing models, whereby they can select only the content that their users read.

Additionally, big deals frequently are priced in a way that is hard to undo or to understand. For example, the University of Minnesota library

---

[3]The information in this section is based on the remarks of Wendy Lougee, director of the University of Minnesota Library.

worked with Elsevier Science to back out of its big deal and found that, because of current electronic and print pricing structures established by this publisher, reducing the subscription list from 750 titles to 650 titles and moving to electronic-only would result in a higher per-title cost.

A second sobering event has been the demise of some of the industry intermediaries such as the subscription agents. One major serial vendor declared bankruptcy, leaving unpaid publisher debts reported to be some $73 million—money that had already been collected from library customers and who ran the risk of not receiving their paid subscriptions for 2003. In the end, most publishers agreed to "grace" the libraries' subscriptions, but at a huge loss to their organizations.

The likelihood of increased revenues for libraries in the near term (particularly increases that match inflation in journal prices) is low. A recent informal survey conducted within the Association of Research Libraries suggested that nearly half of respondents expected cuts in some areas and the prospects were high for further budget reductions in the coming fiscal year. Library budgets, a major source of revenue for publishers, are obviously stressed. The volatility in the publisher marketplace will probably continue, as will the push from the library community for the more finely grained models that allow them to make some choices.

Implications of Changes in Journal Format and Content

Early usage data indicate that much more use results from electronic content, which is available to licensed users anytime and anywhere. Recent studies of university users nationwide have revealed an overwhelming preference for electronic format. In such surveys, nearly half of all faculty in most disciplines reported they use online materials for the majority of their work.

Yet interestingly, despite that preference, other studies of perceptions of convenience and ease of use show a dramatic gap in how the library performs in delivering electronic content. Users cite evidence of their inability to manage such content, to navigate it well, or to deal with the myriad different distribution platforms and channels.

In addition, there is a subtle shift from our concept of publication as product to the notion of publication as process.[4] There are a number of

---

[4]See discussion in the chapter on "What Constitutes a Publication in the Digital Environment?"

examples where online discussions really have the form of being an actual publication. For libraries, which are in the business of managing copyrighted, fixed works, that presents a real challenge. Dynamic "publications" pose a challenge, too, to the STM publishing sector in terms of pricing. How should a publisher develop models that support publications that are not fixed or well bounded?

The Changing Role and Influence of Libraries

A recent Morgan Stanley report[5] suggests the potential for reduced operating costs for libraries—no periodical check-in, no binding, no claiming. However, the necessary infrastructure to support the investment in electronic content, to federate it appropriately, to ensure its longevity, and to archive it, requires greatly increased expenditures on the library side. Any subscription savings will be needed to support additional electronic infrastructure. It is critical to focus community attention on issues of infrastructure, interoperability, and the kinds of protocols that will allow that federation to happen.

Libraries have a role in seeding and supporting alternative, competitive approaches to electronic publishing. Librarians understand content, its use, and the users. There are many examples of libraries actively engaging with new types of STM journal publishing. A growing number of institutional libraries, such as Cornell and Michigan, are starting incubator and production services to help small publishers move to electronic publishing. These projects represent a move away from libraries' traditional role of providing access to information toward facilitating production of information, and it may help libraries reconceive the relative position they have long held in the STM information sector.

THE COMMERCIAL SUBSCRIPTION-BASED MODEL[6]

One fairly typical example of a commercial STM publisher based on a reader-pays model is John Wiley & Sons, Inc., which is a global, indepen-

---

[5]Morgan Stanley Industry Report. 2002. Scientific Publishing: Knowledge Is Power.

[6]This section is based on the remarks of panel participant Brian Crawford, vice president and general manager, Life and Medical Sciences, John Wiley & Sons, Inc.

dent publisher established in 1807. Wiley's three major areas of publishing today include STM journals, higher education materials, and professional and trade information. Theirs is a fairly diversified publishing portfolio, with about $1 billion in annual revenues. Wiley publishes 400 STM journals online on its *InterScience* platform, which was established in 1997. This resource now contains about 2 million pages of information.

Because Wiley uses a customer- or reader-pays model of delivering STM journal content, the publisher places a great emphasis on sales staff worldwide—staff that it did not use in the conventional print environment, where it was marketing via direct promotion to scientists and libraries. As Wiley embarked upon the development of licenses for its electronic journals, it did so very much in consultation with its major institutional customers. This resulted in flexible sales options. The company did not want to make an early commitment to one preferred option early in its electronic publishing, so it provided a menu of print and online options.

Many not-for-profit organizations are still in the early stages of developing bundled journal licenses for their institutional customers, whereas for-profit publishers have undertaken much more aggressive licensing in recent years, as noted above. Wiley currently uses what it calls "Basic Access" and "Enhanced Access" licenses. The basic access license offers title-by-title access, with some concurrent user restrictions. The basic access option is most often suitable for the smaller institution or department. The enhanced access license does not require that an institution subscribe to all Wiley titles that are available electronically. The institution can choose, but it is establishing a license for a larger body of work, with no concurrent user restrictions and with additional benefits such as negotiated price caps. Wiley's business model also emphasizes direct relationships with its customers.

Finally, Wiley constantly invests in new features and enhancements for its electronic publications. Some of the most recent ones include content alerts to apprise its audience of what is being published, delivery of content to mobile edition platforms, and publishing online in advance of print publication.

Reasons That Wiley Will Not Use the Author-Pays, Open-Access Model

There are several reasons why Wiley has not elected to use the author-pays, open-access route. Most of Wiley's professional society partners are quite wary of any economic system that tends to favor author payment for publication. Given that the bedrock of scientific communication is peer-reviewed journal literature, Wiley's view is that any system that charges the author or a sponsor of the author in order for that author to be published is going to favor the author's desire to become published.

What has evolved in scientific communication, for the most part, is a system where the reader pays, or an agent for the reader is paying, because that naturally introduces an objective filter for the validity and the value of the work. One should not assess the value of information on the basis of the cost to produce it. Instead, the value of the information is its value as a tool, as a productivity multiplier in society. Wiley, therefore, has very much supported a customer-pays model for the business, because the company believes it ultimately enhances the value of scientific information for those who should value it most.

Nevertheless, Wiley does offer free access, from the time of publication, to developing countries. For its biomedical journals, it provides free or inexpensive access through the Health InterNetwork Access to Research Initiative (HINARI) of the World Health Organization (WHO). HINARI puts content into the hands of investigators in parts of the world who truly cannot afford such information.

THE OPEN-ACCESS, AUTHOR-PAYS MODEL[7]

The Public Library of Science (PLoS) is a new scientific publisher with an open-access business model, which has also been called a midwife model.

Reconceptualizing the STM Publishing Business Model on the Internet

Before the Internet, there was no choice but to charge users of scientific publications, because the most efficient way of making them available was by distribution in the print format. In the print environment, every potential user represented an incremental expense for the publisher, and any business model that did not take that into account was doomed to

---

[7]This section is based on the remarks of panel participant Patrick Brown, professor of biochemistry at Stanford University and a cofounder of the Public Library of Science (PLoS). For additional information about the PLoS, see http://www.plos.org/index.html.

economic failure. The print system thus had the consequence of limiting distribution only to individuals and institutions that were willing and able to pay, and this system was accepted as a necessary evil. With the advent of the Internet, however, that model is no longer necessary.

The traditional business model has also had another indirect consequence that has been subtle but equally unfortunate: It is based on selling research articles. Some scientific publishers can assert that the content of their journals is valuable property that they own, control, and sell for a profit. This is a barrier to open scientific communication that now needs to be reconsidered.

The worldwide spread of the Internet now leads to fundamental and positive change in the economics of scientific publication, as well as the technical means of distribution. The change makes possible the realization of Jefferson's ideal of the infinite, free dissemination of scientific ideas and discoveries. What had been an impossible ideal in the pre-Internet era—to make the published information an open public resource—is now possible, because the cost to the publisher no longer scales with the number of copies produced or with the number of potential readers of a publication. Accordingly, users are not restricted to a business model that charges per access or per copy. In fact, we see that a business model that restricts the distribution and use of the published work is working against the interest of science and society. The economic model of print has become unnecessary, anachronistic, and inefficient and now stands in the way of the ideal of open and free dissemination. If we do not need to charge readers for access, then we should not charge for it.

"Open Access" Defined

An open-access publication is one that meets two conditions. The first is that the copyright holder (either the author or the publisher, if the copyright has been transferred to the publisher) grants to the public a free, irrevocable, perpetual right of access to, and a license to copy, distribute, perform, and display the work, and to make and distribute derivative works in any medium for any purpose.

The second condition is providing readers with open access to the work. Authors or publishers achieve open access by making a complete version of the article and all supplemental materials available in some suitable standard electronic format, deposited immediately upon publication in at least one internationally recognized, independent online repository that is committed to open access. One well-known example of such an open access archive is PubMed Central, maintained by the National Library of Medicine.

Advantages of the Open-Access Approach for Science

The practical advantages of true open access are already very familiar to many researchers in the life sciences through two longstanding, amazingly successful open-access experiments—GenBank and the Protein Data Bank. The success of the genome project, which is generally considered to be one of the great scientific achievements of recent times, is due in no small part to the fact that the world's entire library of published DNA sequences has been an open-access public resource for the past 20 years. If the sequences could be obtained only in the way that traditionally published work can be obtained, that is, one article at a time under conditions set by the publisher, there would be no genome project. The great value of genome sequences would be enormously diminished.

More significant is the fact that open access is available for every new sequence, which can then be compared to every other sequence that has ever been published. The fact that the entire body of sequences can be downloaded, manipulated by anyone, and used as a raw material for a creative work has led thousands of individual investigators to take up the challenge of developing new data-mining tools. It is such tools and the new databases that incorporate sequences, enriched by linking them to other information, that have made the genome project the success that it is today. By adapting the genome model of open access to the publication of scientific literature, we could see a similar flowering of new, investigator-initiated research and creative, value-adding work.

Open Access Supported by the Author-Pays Business Model

Unlike the subscription-based model, the PLoS plans to charge the costs of publication to authors and their sponsors. From the standpoint of business logic, this is by far the simplest and most natural model. It is natural, because the cost of online publication is scaled to the number of articles, not the number of readers. It also makes sense from the standpoint of institutions that pay for research. Their mission is to promote the production and dissemination of useful knowledge. From that perspective, publication is inseparable from the research that they fund. The PLoS initially plans to charge about $1,500 per published article, with no charge to authors who cannot afford to pay.

Research sponsors should welcome this model, because for a fraction of 1 percent of the cost of the research itself, the results can be made available to all readers, not just to the fortunate few who are at the lead research institutions (i.e., institutions that can afford to pay for site licenses).

In the short term, of course, open access will generate an incremental expense, but in the long term, once the scientific community has made the transition to open-access publication, there will actually be savings to the major research sponsors because, after all, they are ultimately the ones who pay for library and even individual subscriptions. The Howard Hughes Medical Institute (HHMI) is one of the largest funders of scientific research in the life sciences, and it has already endorsed this model. The HHMI has agreed to provide budget supplements to its investigators, specifically to cover author charges for open-access publications.

The PLoS will also produce printed editions of its journals for sale to institutions or individual subscribers at a price intended to recover only the cost of printing and distribution, everything downstream of producing the published digital document. The print subscription is estimated to cost approximately $160 a year. There will be no cross-subsidies between the open-access online publication and the break-even print publication operation.

ISSUES RAISED IN THE DISCUSSION

Relative Merits of the User-Pays and Author-Pays Models

A number of points and counterpoints were made in favor of both types of models, in addition to those raised during the speaker presentations, summarized above. The following arguments were presented in favor of the user-pays model and against the open-access model, in which the author or the institution funding the research pays.

Devaluing the overall utility of the information. There is an inherent bias in a system where the author pays, which ultimately would devalue that information in terms of its overall utility as a productivity multiplier. There are two reasons for this: The first is that a less selective filter would be imposed. And second, is that an economic system where the author pays is naturally going to favor the author. That means that any entity wanting to make decisions about that work needs to impose yet another filter, at some cost, in order to determine how that work stacks up against others. Most scientific publishers would say that selectivity is probably not exercised all that often by authors, because they seek to get their work out. It is necessary to optimize two variables. One variable is the dissemination, and the other is the filtering. The author-pays model moves the filter boundary, whereas publishers also have an interest in disseminating the work as broadly as possible. It is a matter of getting the balance right between those two considerations.

The difficulty of convincing research funders to subsidize authors' page charges. According to statistics presented by Donald King of University of Pittsburgh, roughly one-third of STM journal articles are funded by the federal agencies, about a third of them are funded exclusively within the universities, and the rest by industry. How will the funders of these authors be convinced that they need to pay an additional page-charge fee of $1,500 above the money that they are already paying the authors to prepare those articles? It is necessary to have different kinds of arguments for those three different constituencies. In doing that, it also may be necessary to go further in trying to understand the funding priorities and budget profiles in each sector.

Organizational missions and market forces as determining factors for business models. One aspect about the discussion of business models that has not been adequately discussed is the question of the mission of the organization that is doing the publishing. That does have a significant effect on the business model that is selected. For example, various societies make use of the page charges as a way of supporting their member subscriptions, keeping the member rates low, so that they can provide more benefits to their members. A commercial publisher or a university press cannot really charge page charges, unless it is a journal they are publishing on behalf of a society, because that is not viewed as appropriate. It would be seen as gouging. Color page charges might be acceptable, because color printing constitutes an additional direct cost for the publisher.

One should not look at the customer-pays model, as opposed to the open-access model that has been proposed as an author-supported model, in absolute terms. Indeed, many professional societies have had a hybrid model, where they have benefited from subscriptions at the same time that they have used author page charges and other subventions to help support their publishing programs and to keep the costs to the customer down. That kind of hybrid model has been determined over time by market forces.

The subvention of publishing costs by the payment of page charges and subvention for carrying color reproduction charges are examples of how some organizations have found an effective balance. They have done this even in a user-pays model, implementing certain charges that are passed on to the author, where the needs of the author are seen as unique, and something that the author would want to pay in order to benefit from a service. Most organizations, however, do not make that a criterion for a decision of acceptance or rejection; it is merely a matter of presentation of the work.

The factor of author selectivity favors a reader-pays model. Finally, the value and the future of any publication are going to be determined mainly by what authors want to do, where authors want to publish. For example, the *New England Journal of Medicine* is open to all authors; 4,000 scientific articles are submitted to the journal each year, with no charge for submissions and no page charges if the paper is accepted. The role of a biomedical or scientific journal is to be critical and selective. That is in part why authors want to publish there. If authors or their sponsors have to pay, is this selectivity going to be compromised? How is the PLoS going to exercise the functions of peer review and of being selective? Is that part of the PLoS model?

Arguments in favor of the author-pays, open-access model and against the reader-pays model included the following:

No correlation between a reader-pays model and the maintenance of high-quality standards. There have been a number of studies that have looked at the relationship between the price and measures of quality of scientific journals—citations, their assessment by peers in the field, and so forth. They have found, overall, a dramatic negative correlation between price and quality. The notion that users' paying for journals somehow upholds high-quality standards is not supported by the data.

Every scientist, at least in the life sciences, knows that most of the journals that are regarded as third-tier journals of last resort, but that publish 99 percent of the articles, have no author charges but very high subscription costs, whereas the premier journals typically wind up charging authors in color charges and page charges something on the order of $1,000 or more.

The authors' overriding interest to be read and to enhance their professional reputation benefits from open access. The factor that serves to maintain the quality of the work that authors submit for publication is not that they think it is extremely difficult to get a paper published. One can always find a mediocre journal that will accept just about anything. Rather, it is that the authors know that sooner or later their peers are going to read what they write, and their reputation depends on it being good. That is ultimately what determines their career advancement, their status in the field, and so forth.

The point that publishers would favor the author by charging the author is absolutely right. The public good that is produced by scientific research is very special. In contrast with other public goods, the value of scientific public goods increases with use, and it is specifically in the author's interest. It is very much a part of the motivation of the author of a scientific paper to have access to that paper as widespread as possible, more and more used, thereby enhancing the value to the community, as well as the interests of the author.

The motivations and interest of all sectors in scientific research should be supportive of an author-pays, open-access approach. The interests are similar for government, academia, and industry, namely, the motivation to support the research and to encourage the authors to publish it. The motivation to pay an increment of less than 1 percent of the total research cost to make it much more valuable to the people who are supposed to be served by it is presumably the same for all three sectors that sponsor research.

Author selectivity not dependent on a reader-pays model. The journals most attractive to authors tend to be the ones in which they have the least chance of having their papers accepted. The PLoS certainly has factored that into its development strategy. It intends to be very selective from the beginning; it is selective not just on the basis of whether the article is good enough to be published somewhere else, but in selecting papers that are likely to be of interest to a very wide audience, precisely because the PLoS considers that this is going to be important in terms of developing the journal identity and as a magnet for submissions.

Page charges not a disincentive to authors. If we go to a model where the institutions are covering page charges as an essential part of research, it will make it even less of a disincentive to authors. Established publications may not need to worry about converting to a system based on author page charges, and the resulting open access would be better for the community they are supposed to be serving.

The Effect of Different Publishing Business Models on the Long-Term Preservation of Digital Journals

The focus of publishers in electronic publishing is mainly on the actual production of the work and its dissemination on the Web. They generally have not addressed the important issues that deal with long-term preservation, including the integrity of the information and its migration to new platforms. Reed Elsevier has begun an innovative project with the Royal Library in the Netherlands, where they are looking at how to address those kinds of long-term preservation problems. The solutions will be costly, and the costs will need to be shared between publishers and libraries, ideally with support from governments and other institutions.

The PLoS definition of open access includes the immediate deposition of the publication with some organization that is committed to long-term access. There is no certainty, however, that organizations will actually follow through on this commitment. One of the advantages of print publication is that with enough copies produced, very long-term preservation is more or less guaranteed, which is not always the case with electronic publications.

The PLoS nonetheless does have plans for providing archival stability for its information. What is important, however, is not only that somewhere there exists a permanently preserved copy of the information, but that it is permanently openly available to everyone. Of course, one would be hard put to find a more trusted and trustworthy archival repository than the National Library of Medicine, which has agreed to archive PLoS publications, but the fact that the information is also going to be freely and widely distributed, and that it will exist in many institutional servers, provides another measure of reassurance. In the view of some observers, it would be a mistake for institutional repositories to agree to take on the job of archiving information without requiring the publishers to grant unrestricted open-access rights.

One example of a professional society's approach to the archiving problem is the American Physical Society (APS), which has put all of its content going back to 1893 online. It is all linked and searchable, and PDFs are also available. Current content is added as it becomes available. There will be costs in changing the format of this in the future. The society also recognizes the concerns of libraries and of the entire scientific community that this historical record might be lost.

What would happen if the APS were to go under? There is a full mirror site of the entire archive that is already accessible and tried at Cornell University, which is where *The Physical Review* originated back in 1893. The journals are also deposited at the Library of Congress. If the society is terminated, it has an agreement with the library that these holdings will be put in the public domain, freely available to anyone. This is the society's primary effort for ensuring continued availability of its archived publications.

There appears to be a clear distinction between the perspective of the professional society as a publisher and the private commercial publisher. The professional society has a very strong vested interested in archiving, whereas the private publisher really has none.

Issues in Transitioning to the Open-Access Model

The APS also has an open-access model for one of its journals that it started 5 years ago—*Physical Review Special Topics, Accelerators and Beams.* It is a small journal, but it is available completely without access barriers. It is not, in fact, free because it does cost something to publish it. The costs are recovered through the sponsorship of 10 large particle accelerator laboratories around the world. It is also cosponsored by the European Physical Society's Interdisciplinary Group on Accelerators. At this point it is only a limited experiment, however, because the APS has other expenses that it must recover, and it has to see whether these things work or not. The society cannot afford to bet its entire future on the open-access approach, because it still needs to recover its costs.

The APS did propose that the society would like to put everything online without access barriers. It could do this right now if every organization that now subscribes to its journals would make those subscriptions sponsorships, providing enough to recover its costs. Then it could be opened to everyone else. The risk would be that institutions might be tempted to decide that since all journals are available, they no longer have to pay anymore. The libraries would love that, but then the publisher would go under.

The question thus arises as to whether there are any paths or way stations between full open access and the current subscription-based model. One possible approach might be for a publisher to give the author an option for a surcharge. The author could pay extra to have immediate open access to his or her article, whereas all the other articles of authors who did not pay the surcharge would be free after 6 months. The amount of the surcharge, however, must not be so high as to be a disincentive. Nonetheless, there is clearly difficulty making that transition. The PLoS had to get a large grant from the Moore Foundation to buffer the financial risk for its experiment. Without the grant, the PLoS could not have been started.

Publishers also could ask the institutional subscribers that now provide most of the revenues—mostly academic institutions that probably accept the philosophy that journals serve the public interest—to continue to pay their subscription fees at the current rate for some interval of years to be specified, during which time the publisher would make the transition to open access. With such a multiyear commitment of support, the publisher would have a stable revenue source that is not put at risk by making that transition. It can try to make the transition, at the end of which time it can determine whether it looks like it is going to be a self-sustaining model. One would hope that the current subscribing institutions would not take the low road and try to undermine the process by free riding and saving themselves a little money. They should see that it is in their own best interest in the long run to encourage the open-access approach.

Such an arrangement is not unlike what Ohio Link does, but that initiative came up with some extra money that enabled the organization to open its journals to the institutions in the entire state of Ohio. That is the kind of catalyst that may be needed: a bit of extra money.

It might be best to view this approach as transitional, because ultimately the sensible thing would be for the research sponsors to cover the publication costs as an essential part of their mission of promoting and disseminating research. In the short term, it is probably necessary to catalyze the process.

The problem with the temporary approach, however, is what happens if it does not work? It is very hard to get people to resubscribe. Once a library has given up the subscription and used the money somewhere else, resubscribing becomes viewed as a new acquisition.

To counter this problem, the sponsoring institutions could be provided an incentive to maintain support. For example, if a university provides $5,000, or whatever the amount may be, then everyone from that university has open-access publishing rights in the journal. This would make it a competitive advantage for the university to offer this. If you treat it as kind of a credit pool that could be drawn on by authors from the subscribing institution, then even in the open-access model, they are getting some special benefits beyond what the nonpaying institutions are getting. It becomes an added incentive.

Advertising Revenues in Electronic Publishing

Nearly 50 percent of one publisher's total costs are attributable to editorial, peer-review, and production processes, items other than printing, paper, and distribution. Although many STM journal publishers have no advertising revenues, advertising for some print journals underwrites more than 50 percent of the total cost of operation. For such journals, the electronic-only model, assuming it had less advertising, might save on some costs, but would actually result in a major lost source of revenue.

For journals with broad member or individual circulation, most advertisers still seek print as their means of reaching that audience. They are not yet ready to move away from print advertising to online-only in scholarly publications, although this may not be the case with other consumer-oriented publications. It would be very risky to go to an online-only strategy for STM publications if the publisher currently relies on such advertising revenue.

Sponsor-Supported Open-Access Model

Many researchers like the open-access model because it is the sponsor of the research who shares with the author the interest in having the product disseminated as widely as possible.

For example, when the National Institutes of Health (NIH) supports most of its research, it is because it will be published and made into a public good. To fulfill the goals of supporting the research at all, the sponsor generally carries a responsibility to see that the material gets published. For this reason, the HHMI model that supports the PLoS is a very good one, and many scientists who publish research would like to see that kind of approach propagate. The question that was raised earlier in this regard still remains, however: Is there a way to persuade the less well funded science agencies to take on the responsibility to pay extra for publication costs? This would be in contrast to the present model, which puts the author payment responsibility into a grant in which the author has discretion whether to spend that money for publication, or to spend it for support of another graduate student or some other research cost.

One can make a strong argument that the publication costs amount to less than 1 percent of the research costs (in biomedical research, at least). If that is true more broadly across the different disciplines, if you had to take a 1 percent cut from other aspects of the budget to do it, it is a plausible argument that the return on that investment would be extremely high, as opposed to the 1 percent cut from other areas, because all the grantees and all the research that an agency is funding would be providing much freer access to a more extensive body of information. The purpose of funding of research is not just to serve the immediate community of the grantees, it is the wider scientific community and the general public that should be much better served by the information.

At the same time, according to Patrick Brown, when this issue has been raised with respect to NIH funding, many NIH grantees have objected if this were to come at the expense of a 1 percent cut in research funding. They argue that there is not enough research funding to go around already. So clearly it is a controversial proposal.

4

Legal Issues in the Production, Dissemination, and Use of the Journal Literature

COPYRIGHT BASICS: OWNERSHIP AND RIGHTS[1]

Basic Rights Conferred by Copyright Law

The very broad construct of copyright protects works of authorship and covers scientific works of all kinds in every medium of expression, such as printed journals or computer software. Although copyright protects the tangible expression of the author's created work, it does not protect ideas or facts, such as genome sequences. It only protects the work that describes or expresses the results and analysis of that information.

Copyright confers upon the author or copyright owner certain exclusive rights, which include the right to reproduce the work in any medium, including digital; the right to make a derivative work, that is, the right to adapt the work, including updating or making further works based on the first work; the right to distribute the work in copies, including digital copies; and a right that is becoming of increasing importance on the Internet, the right to publicly display the work.

The exclusive rights under copyright are not absolute. There is the well-known fair use provision, Section 107 of the Copyright Act, which

---

[1]This section is based on the presentation by Jane Ginsburg, Morton L. Janklow Professor of Literary and Artistic Property Law at Columbia Law School.

modifies the exclusive rights of owners. Copyright is limited in time as well. The term of copyright in the United States is now the life of the author plus 70 years. That is a very long time, and it is the same term in the European Union.

Ownership of Copyright

Copyright ownership vests in the author of the work. However, according to copyright law, the employer of a work for hire is considered to be the statutory owner of such a work. A work for hire is one prepared by an employee within the scope of his or her employment.

As a practical matter, the question of who owns professorial or academic copyright did not arise until relatively recently. That has changed in the past few years, in part because of the rising importance of software. As a result, some universities began to lay claim to software, not only software written by staff but also by professors. Universities traditionally had distinguished between works by administrators or staff, which they defined as works for hire, and works by professors, which were not considered works for hire.

Additionally, given the continuing escalation of journal subscription prices, some universities believed that by owning the copyrights on articles, they could bargain better with the publishers. Finally, the most recent flurry of copyright ownership policies was precipitated by distance education on the Internet.

About two-thirds of the universities assert that faculty own the works that they create, subject to a number of fairly typical default-shifting conditions. First, if the university has invested substantial resources in the creation of the work over and above resources commonly made available to faculty, then the university is likely to assert ownership in that work.

Second, many universities define a category of institutional works that border on the administrative, such as courses that are uniform throughout the university. Those are frequently the objects of university assertion of copyright.

Third, many universities have special provisions for "sponsored" research (by an outside entity). If a condition of the outside entity is that the university should own the copyright, then the university will assert ownership of the resulting work.

Fourth, a number of schools will assert copyright ownership if the work qualifies as "technology" or software.

When the default position is that the university owns faculty work, there is generally a provision listing the circumstances under which the university will forbear from asserting copyright in faculty-created works.

Regardless of who owns copyright in a university, most policies limit the university's or faculty member's exercise of copyright. The most typical constraint on faculty in the exercise of copyright is that the faculty member will be asked to give the university a royalty-free license to make nonprofit university educational use of the faculty member's work product. Often that license applies even after the faculty member has left the university.

If the university owns copyright, whether by default or because one of the conditions for transferring ownership back to the university applies, the most typical constraint on the university's exercise of copyright is to allow faculty noncommercial use rights. Some universities will not commercialize a work without obtaining the reasonable consent of the faculty member in advance. Furthermore, many universities provide that with respect to course materials, even when the copyright is owned by the university, if the faculty member leaves and goes to teach somewhere else, he or she can take those materials to the next place.

Only one university, Columbia, provides that even when the university owns copyright in a work, it should respond favorably to the creator's request to the university to make the work freely available.

Rights That a Faculty Member Gives Up To Be Published

Assuming that the faculty member does own copyright to a scientific work, what does he or she have to give up in order to be published in a journal? Interestingly, there is not much difference in the terms between the societies and the commercial publishers.

Most publisher agreements do, nonetheless, provide for authors to use and reuse their articles, notably for their own nonprofit use in further works or in teaching. Where there is considerable divergence is with respect to electronic rights. Almost all STM journals publish both in print and in electronic form. Depending on the contract, the author of the article will be permitted to post an abstract of the article with a link to the publisher site, or she may be allowed to post a preprint of the article on her own Web page, or on some other preprint site, but she may not be able to post the final version of the text as edited by the journal. Another variation permits posting of the article but only on restricted access sites. So, what authors can do with their own articles falls well short of open access.

One might think these policies of publishers would create a lot of public opposition by the authors, but they have not. Why not? The simple and completely unlegal answer is that professors in practice largely ignore these contractual limitations, and they boldly post full text regardless of what their contract says. The authors are basically betting that the STM publishers will not actually enforce that contract against them.

Nonetheless, copyright does matter to authors. There is a significant psychological factor in being the owner of the copyright in your work. The universities that take the work-for-hire position—to the extent that their faculty know about it—risk antagonizing their faculty members. There is something very deep seated, even if one is disposed favorably to open access, in being considered the author and copyright owner of your work. Whether or not copyright will matter to publishers in the future, it will most likely continue to matter to authors.

LICENSING[2]

Licenses are not entirely new, but are a more recent method for distributing STM journals and other digital information. Licenses, or contracts, are private, negotiable agreements. They are specific and very tailored, and that specificity can be very reassuring.

Licenses can restrict or they can expand rights that would be granted under copyright law. In that sense, a license is a neutral instrument controlled by the parties to the agreement. Licenses used to be regarded as entirely controlled by publishers, at least in the library community, but this is no longer true in all cases. Libraries, working with publishers in the license environment, have in fact made an electronic market possible.

Major Licensing Issues for Libraries

Not all licenses have been fair or negotiable for libraries. In fact, 10 or 15 years ago, licenses offered by publishers or vendors to libraries were often unacceptable. Some problem areas have been resolved, but others persist. The most difficult areas in which to reach agreement have been the terms for interlibrary loans and guarantees of perpetual access and archiving.

---

[2]This section is based on the presentation by Ann Okerson, associate university librarian for Collections and Technical Services at Yale University Library.

Interlibrary loan is a relatively long established practice whereby a user in an institution that does not have a book or article can request it from another library. Articles are generally photocopied and sent to the requester by fax or by mail. Now that articles are available online, one might expect that interlibrary loans should be able to take place with the supplying library transmitting the article electronically to the requesting user. In fact, very few publisher licenses permit this type of transmission. Most require that articles be printed from the e-version and then forwarded, or they do not permit any interlibrary loan from the electronic version at all.

When the interlibrary loan from the e-version is only permitted after it is printed out, the main consequence is that delivery is more cumbersome than it otherwise would be. Where the interlibrary loan for the e-version is not permitted at all, then a serious degradation of access ensues in the electronic environment, especially as libraries move to electronic-only subscriptions.

The concern about archiving and perpetual access in licenses is somewhat different from this, but it is nonetheless a great concern. Research libraries have indeed attempted to hold onto print subscriptions for their archival and preservation value and to adopt electronic subscriptions to journals for reasons of service and functionality. However, the economic climate is such that many libraries, for lack of resources to support both print and electronic subscriptions, are beginning to drop the print. This is happening even in the biggest research libraries.

Libraries have always assumed that the material they pay for will last indefinitely, and users will be able to have long-term access to it. The movement to electronic-only suggests two requirements. One is that repositories of electronic journals must be robust, even though currently they are less than optimal. The other is that libraries licensing journals on behalf of their users will want continued access to those e-resources even after their paid access to a given journal stops for some reason.

The majority of contracts for electronic journals now do provide language for continued access, often in perpetuity, but not all do. Furthermore, the contracts are very weak because of the inability of most publishers to follow through technologically or in a business sense on the promises that are made about the archiving provisions.

These major flaws in archiving provisions will be resolved only upon significant discussion and investment by all the stakeholders with regard to archiving responsibilities, costs, and technologies. Meanwhile, as libraries are now canceling paper subscriptions, the official scientific record is at some considerable risk.

Licenses That Serve Author Interests Better

Scientists clearly want their articles to be widely available. Many scientists today might like to distribute their own articles on their Web sites, their university sites, their lab sites, and e-print sites, in addition to the formal peer-reviewed journals. Some publishers permit this, even though copyright has been transferred to them, but others do not. In most cases, publishers ask scientists to transfer their copyright, and scientists are accustomed to make such transfers and sign those papers.

How can authors get out of this quandary? There are two ways of doing this. A good way is for authors to sign a copyright transfer that also permits them to redistribute their own work. The advantage of this to authors is that it keeps them out of downstream copyright management problems.

An even better way, perhaps, is for the author to retain copyright, while licensing the publisher to deploy the work in all the ways that the publisher needs for effective publication and dissemination. In this case, the author will have continued responsibility for managing his or her copyrights, but will also have broad flexibility as owner of the work.

ECONOMIC AND NON-ECONOMIC REWARDS TO AUTHORS: THE SOCIAL SCIENCE RESEARCH NETWORK EXAMPLE[3]

The Social Science Research Network (SSRN) was not founded to make profits, but as a way to change or to make efficient the distribution of work in the social sciences. The SSRN posts on its Web site[4] abstracts of full-text, nonrefereed working papers, as well as full-text, refereed articles from journals of established publishers who want to have access to the community that has been created around the Web site.

The SSRN has created some measures of popularity. Although it does not referee anything, SSRN's rules are that it will post material that is part of the worldwide scientific discourse in the field for which it is intended. For each field and each journal, SSRN creates "top 10" lists, based on use. The top paper has had 30,257 downloads. Posting articles electronically results in an unbelievable amount of attention for papers, at least compared with normal usage of a printed journal.

---

[3]This section is based on the presentation by Michael Jensen, managing director of the Organizational Strategy Practice at the Monitor Company, and Jesse Isidor Straus Professor of Business Administration emeritus at Harvard Business School.

[4]For additional information on SSRN, see its Web site at http://www.ssrn.com/.

Every author in the system has an author page. Wherever the author's name shows up on the SSRN Web site, it is hyperlinked to a stable URL that provides full contact information for the author. The author page also provides a list of all the author's papers on the SSRN site, and these papers are hyperlinked and available for downloading. The author page reports the total number of downloads from all of the author's documents, the articles' ranks, and the downloads for each one of them.

The great value of SSRN in developing those top 10 lists is that it has devised an infinite number of carefully gauged categories. Almost every SSRN author at one point in his or her career is a "top 10" author. That author receives congratulatory e-mail from the SSRN editors, even for an article in a narrowly defined category. It can be very gratifying. Authors can look at their Web page for the number of downloads of an article in any given week. That kind of feedback is one of the reasons that the SSRN gets a high level of participation, once authors know about it. Even though SSRN does not offer the kind of prestige credentialing that is provided by peer review, there is a certain amount of validation from it.

ISSUES RAISED IN THE DISCUSSION

Problems Encountered in the Transfers of Copyrights

Faculty may not even know who owns the copyright in their university. If an article is a work made for hire, then the author may in fact be selling the Brooklyn Bridge when signing the publishing contract. However, most universities have written copyright policies and require their faculty to sign a special agreement, or the faculty member's employment agreement incorporates the copyright policy by reference. If the university does not assert copyright or grants back the copyright in traditional academic scholarship to the professors, that, of course, transfers the copyright to the professors, and then the professors have something to give to the publishers. It is true, however, that the ambiguity about the actual status of faculty writing potentially affects a lot of publishing contracts as well.

Problems with the University Work-for-Hire Approach to Academic Publications

Several more issues may be highlighted with regard to academic work for hire. Many academics move from one institution to another. In that case, which institution owns the copyright? Also, the principle of academic freedom suggests that the professor creates concepts, develops them independently, and the work is not done at the behest of the university. All the university really requires is that the professor be productive, so how can the university claim copyright?

Burdens for Small Publishers in Developing Countries from Licensing and Restricting Access

The publishers associated with Bioline International are all typically very small, nonprofit scholarly journals from developing countries. Their major objective is making their materials visible, and by electronic means they hope to accomplish that goal.

Another obstacle has been getting their journals into libraries. Big libraries have an advantage because they have the staff to negotiate a license. Also, the licensing process favors big publishers, who have sales and legal staff to handle the negotiations and contracts. Bioline International has a permanent staff of one and a half, so they do not have the know-how or the time to negotiate with libraries or to work out a system to get their material to the right outlets.

For small publishers with only several hundred members and very limited distribution, open access is the only way to go. The goal of the not-for-profit publisher is to fulfill the mission of the organization, that is, to make available the information from research, rather than to make a profit. There should be other ways to recover costs, rather than to get stuck in the subscription model as a way to pay for publication. A related lesson is that to control access, small publishers will spend more money than they are able to recoup from their subscriptions. Bioline International needs to measure return on investment in publication in terms of the readership and the impact, rather than the revenue that it can generate. It hopes to convince funding agencies that support these journals that they are spending their money in a useful way.

5

What Is Publishing in the Future?

The digital revolution is changing the traditional processes of many knowledge-intensive activities, in particular the publishing or scholarly communication processes, and it is offering alternatives both to how the various stages of these processes are conducted and who does them. The various functions—whether metadata creation, credentialing review, or long-term stewardship—can be separated or disaggregated, and players different from those who traditionally have carried out these tasks can, in theory, perform them. Publications can now exist in many intermediate forms, and we are moving toward more of a continuous-flow model, rather than a discrete-batch model.

The raw ingredients—the data, the computational models, the outputs of instruments, the records of deliberation—can be online and accessible by others, and can be used to validate or reproduce results at a deeper level than traditionally has been possible. Third parties—particularly in an open-access, open-archives context—can then add value by harvesting, mining, enriching, and linking selected content from such collections.

The presentations in this session of the symposium identified some of the social processes, specific pilot projects, and the challenges and opportunities that may provide the basis for future "publishing processes," which ultimately may be more holistically integrated into the "knowledge creation process."

IMPLICATIONS OF EMERGING RECOMMENDER AND REPUTATION SYSTEMS[1]

When some people think about changing the current publication process to a more open system, they express concerns that scholarly communication will descend into chaos and that no one will know what documents are worth reading because we will not have the current peer-review process. This idea should be turned on its head. Instead of going without evaluation, there is a potential to have much more evaluation than we currently have in the peer-review process.

Public Feedback

There can be a great deal of public feedback, both before and after whatever is marked as the official publication time, and we can have lots of behavior indicators, not just citation counts. Various Web sites now evaluate different types of products or services and post reviews by individual customers. For example, the reviews at Amazon.com provide both text reviews and numeric ratings.

Even closer to the scientific publishing world is a site called Merlot, which collects teaching resources. Merlot provides for a peer-review process before a publication is included in the collection, but even after material is included, members can add comments. Many other examples of this type of public feedback, both within and outside the STM communities, already exist.

Behavioral Indicators

With behavioral indicators, the researcher does not ask people what they think about something; he or she watches what they do. For example, Amazon.com, in addition to its customer reviews, provides a sales rank for each book. Google uses behavioral metrics of links in its page rank algorithm. The Social Science Research Network uses the download count as a behavioral metric.

---

[1]This section is based on the presentation by Paul Resnick, associate professor at the University of Michigan.

Potential Problems

The existing examples of public feedback approaches and behavioral indicators provide some potential models for developing credentialing processes for scientific publication in the future. There are also some problems that require further examination, however.

An obvious potential problem is "gaming the system." For example, an owner of a Web site can hire consultants to help the site get higher Google rankings. No matter what the system, some people will try to figure out what the scoring metric is and attempt to influence it to boost their success in the rankings. Another problem is eliciting early evaluations. In those systems where there is widespread sharing of evaluations, there is an advantage to being second, to let somebody else figure out whether an article is a good one to read or not. Yet another problem can be herding, where the later evaluators are overly influenced by what the previous evaluators thought.

Experiments to Try in STM Publishing

Some potential experiments are more radical than others. Journal Web sites could publish reviewer comments. The reviewers might be more thorough if they knew their comments were going to be published, even without their names attached. The reviews for rejected articles could be published as well. Such a system could reduce really poor submissions.

Other experiments might try to gather metrics. Projects such as CiteSeer in the computer science area measure citations in real time. Experiments in evaluating the evaluators are needed as well. More attention, greater credit, and rewards need to be given to reviewers for evaluating early, often, and well. Publishers already are complaining about the difficulty of finding good reviewers.

Finally, the thread of the original version of a paper, along with reviewers' comments and authors' revisions or responses to the comments, as well as the journal editor's additional comments and interpretations, could be used as educational material and enrichment. All this could be done either anonymously or by attribution.

PREPRINT SERVERS AND EXTENSIONS TO OTHER FIELDS[2]

The article preprint is a well-known, well-understood concept in the physics community, but is not as well known in other communities. Preprints have a well-understood "buyer beware" connotation in the physics community. They provide a means to get informal, non-peer-review feedback that is weighted very differently by physicists than a formal refereed report. Preprints get an early version of an article out to colleagues to solicit feedback.

The e-Print arXiv in Physics and Other Similar Projects

The e-Print arXiv[3] was created by Paul Ginsparg (then at the Los Alamos National Laboratory) for the high-energy theoretical physics community. Today the archive is hosted at Cornell (where Ginsparg works) and covers 28 or so fields and subfields in physics, with more than 244,000 papers. It has succeeded in large part because Dr. Ginsparg is a physicist. He understood well how that community works and what its needs are. The arXiv clearly has increased communication in its field. It is the dominant method for authors to first register publicly their ideas. It addresses the common interests of a community in a sociologically compatible way.

This approach has spread to other fields, including biology, materials science, social sciences, and computation. Spinoffs have been organized by both universities and government agencies. CogPrints,[4] at the University of Southampton in the United Kingdom, is a well-known preprint system in cognitive science. NCSTRL, the Networked Computer Science Technical Reference Library,[5] was an early effort for harvesting computer science papers and building a federated collection of that literature. The National Aeronautics and Space Administration National Technical Reports Server[6] was a pioneer in bringing together and making available a collection of

---

[2]This section is based on the presentation by Richard Luce, research library director at Los Alamos National Laboratory.

[3]For additional information on the e-Print arXiv, see http://www.arXiv.org/.

[4]For additional information on CogPrints cognitive sciences e-print archive, see http://cogprints.ecs.soton.ac.uk/.

[5]For additional information on the NCSTRL archive, see http://www.ncstrl.org/.

[6]For additional information on NASA's National Technical Reports Server, see http://ntrs.nasa.gov/.

federal reports, both metadata and the full text. PubMed Central is certainly well known in the life sciences community.

Inspiration for the Future

What do these developments mean beyond the physics community, and what new efforts can we foresee? A major question to be addressed is how peer review will work in the preprint and open-access contexts. How should we evaluate, recognize, and reward this scientific work?

A variety of methods are worth considering in a composite approach. Today we use citations as the sole indicator of influence, but the problem is that the citation is only one indicator of influence. A better approach might be to supplement the current system with a multidimensional model to balance bias. An ideal system might have the following elements: citations and co-citations; the semantics, or the content and the meaning of the content in articles to see how they are related; and user behavior, regarding how readers use scientific information online. The latter metrics raise privacy concerns, however.

INSTITUTIONAL REPOSITORIES[7]

The mission statements of many universities proclaim that they are committed not only to generating knowledge but also to disseminating and preserving it. Massachusetts Institute of Technology (MIT) decided recently to initiate two projects that would better implement this mission. OpenCourseWare[8] aims to provide free access online to the primary materials for the university's courses. The DSpace initiative,[9] which is the sister project of OpenCourseWare, is a prepublication archive for MIT's research, supported by an institutional commitment from the MIT libraries. DSpace is meant to be a federation of the intellectual output of some of the world's leading researchers. MIT will not build the whole system for DSpace. Instead, it is creating elements that communicate through open standards, so

---

[7]This section is based on the presentation by Hal Abelson, professor of electrical engineering and computer science at MIT.

[8]For additional information on MIT's OpenCourseWare project, see http://ocw.mit.edu/index.html.

[9]For additional information on the DSpace Federation, see http://www.dspace.org/.

that many users can enter at different places in the value chain and add value in different ways.

Both OpenCourseWare and DSpace are ways that MIT and other universities are asking what their institutional role should be in disseminating and preserving their research output. Why ask this question now?

The answer is that the increasing tendency to proprietize knowledge, to view the output of research as intellectual property, is hostile to traditional academic values and to the public missions of universities. The current research information environment includes high and increasing costs; imposition of arbitrary and inconsistent rules that restrict access and use; impediments to new tools for scholarly research; and risk of monopoly ownership and control of the scientific literature.

The basic "deal," as seen by many in universities, is that the authors, the scientists, give their property away, through copyright transfer, to STM journals. The journals then own this property and all rights to it for the duration of copyright. The publishers take this property and magnanimously grant back to the authors some limited rights. The universities, who might have a stake in ownership transfer, generally retain no rights at all, and the public is not even a factor in this discussion.

It is instructive to list some of the elements that are valuable for promoting the progress of science. They include quality publications and a publication process with integrity; open, extensible indexes of publications; automatic extraction of relevant selections from publications; automatic compilation of publication fragments; static and dynamic links among publications, publication fragments, and primary data; data mining across multiple publications; automatic linking of publications to visualization tools; integration into the semantic web; and hundreds of things no one has thought of yet. OpenCourseWare and DSpace are only two examples of the changing role of universities in using the new technological capabilities to reinforce their public missions and promote the progress of science.

ISSUES RAISED IN THE DISCUSSION

Assessing Journals and Authors

In developing new feedback systems, it is useful to factor in the enormous value that is derived from the literature outside academia. There are two purposes for doing this. One is that it is a better way of assessing journals and authors. The other is that it also will begin to develop a means for the authors to better recognize that their larger audience is outside of their immediate peers.

Indicators versus Metrics

It is better to use the word "indicators," rather than "metrics," because there is no one number that can be used as a measure of quality. There are problems with all measures, including the opportunities for gaming the system. One has to look into all of these and use them as indicators, and it takes a number of indicators to develop a fair measure of quality.

Open versus Confidential Peer Review

Although public commentary is good, nevertheless there is a sort of "Gresham's law of refereeing," whereby bad referees tend to drive out the good ones. The knowledge that a paper is going to be peer reviewed does have an effect on authors.

One way to evaluate the reviewers is to have an editor who chooses them or moderates. One could also develop a system to calibrate reviewers against each other. Unfortunately, this leads to overburdening of the good reviewers, so they are in effect punished for the good work that they do. Some participants believe the peer-review system could be adapted to a more public version. In a public system, one does not necessarily have to give all the lousy reviewers an equal voice.

Constraints of the Print Model on the Electronic Model

Many people are still working in the paradigm of the old paper model of publishing, where there is a lot of prepublication work, because a big print run is needed in order to economize on the costs of distribution. The questions about archival policy, editorial policy, and open access all change completely if one moves to a model of continuous improvement of the materials, or continuous publication, where all peers have an opportunity to adjust the prominence of newly developed pieces. In that model, the world changes completely.

A prerequisite for this kind of innovation is that the materials that are being continually re-evaluated are an open public resource. It is hard to see how this kind of approach would work, as a practical matter, if you still have a system in which every publisher's body of work is in a separate, restricted repository.

Relationship between Open Archives and Traditional Publishers

One can see that repositories such as DSpace have a very valuable role in shaking up the system and in helping to establish or return to better priorities in scholarly publication. There is no inherent hostility between institutional archives and traditional publishers. One can imagine a university holding both the preprint and the final edited version of an article, and the journals providing some kind of authentication and review service.

Preprint servers are repositories at the "lower" layer and provide a platform or an infrastructure on which a whole host of yet to be fully imagined value-added features could be built, some of them by for-profit entities. The goal is to create a more open environment for the primary, or upstream, parts of the value chain and then to encourage scholarly activity on top of that. In any event, of deep concern to the government science agencies and to public institutional repositories is being able to have access to material created with public monies, and to make such information publicly available.

6

What Constitutes a Publication in the Digital Environment?

As we study the question of what constitutes a publication and how the character of publications is changing, we switch our focus from the environmental questions of publication as process to how to author and bind together pieces of authorship into structures like journals.

Publication used to refer to the act of preparing and issuing a document for public distribution. It could also refer to the act of bringing a document to the public's attention. Now, publication means much more. It can refer to a document that is Web-enriched, with links, search capabilities, and potentially other services nested in it. A publication now generates usage data and provides many other functions.

We can approach this from two kinds of perspectives. One is from the individual author's point of view. The practice of science is changing. It is becoming much more data intensive. Simulations are becoming a more important part of some scientific practices. We are seeing the development of community databases that structure and disseminate knowledge alongside traditional publishing. From that perspective, one might usefully ask questions about how people author articles, given the new opportunities technology is making possible. It is clear that articles can be much more than just paper versions by digital means.

The other perspective is that of the journal publisher, regarding the aggregation of these articles, recognizing that the ecology in which journals exist has changed radically. There are all kinds of data repositories. There are live linkages among journals. There is an interweaving of data and authored materials that is becoming very complex.

In this session we look at three innovative examples of publications and some of the issues they raise.

THE SIGNAL TRANSDUCTION KNOWLEDGE ENVIRONMENT[1]

The goal of the Signal Transduction Knowledge Environment (STKE),[2] developed by *Science*, was to move beyond the electronic full-text journal model. The idea was to provide researchers with an online environment that linked together all the different kinds of information they use, not just their journals, so that they could move more easily among them and decrease the time that was required for gathering information, thereby giving them much more time for valuable research and increasing their productivity. The STKE was the first of the knowledge environments hosted at HighWire Press. There are now five.

The STKE has both traditional and nontraditional types of publications and functions. In addition to providing access to the typical journal literature, *Science* tried to create a community environment, to provide an area with tools that relate to that community, and resources that scientists would use. The STKE is trying to create new knowledge and look for ways to explore the network property, the signaling systems that one cannot get from the print product.

The STKE virtual journal has full-text access to articles about signal transduction from 45 different journals. When these journals are placed online by HighWire Press, the STKE uses an algorithm that scans the content and selects the materials related to signaling, which the subscribers to the STKE can then access.

The community-related functions of STKE include letters, discussion forums, and directories. That has been the hardest part to develop.

There already have been some lessons learned from the STKE experiment. The definition of a publication is evolving. Efforts to standardize

---

[1]This section is based on the presentation by Monica Bradford, executive editor of *Science*.

[2]For additional information about *Science*'s STKE, see http://stke.sciencemag.org/.

data input and control vocabularies have been really difficult. Perhaps most important, the reward system is not yet in place for those who are doing this kind of authoring.

PUBLISHING LARGE DATA SETS IN ASTRONOMY— THE VIRTUAL OBSERVATORY[3]

Why is the publishing of large data sets an issue? Scientific data are increasing exponentially, not just in astronomy but in science generally. Astronomers currently have a few hundred terabytes derived from observations made in multiple wavelengths. Very soon, however, they are going to start projects that are going to reobserve the sky every four nights, to look for variable objects in the temporal dimension. At that point, the data will increase to a few petabytes per year.

Astronomy, as most other fields of science, operates under a flat budget, so astronomers spend as much money as they get from their funding sources to build new observational tools and computer equipment to get more data, which they can analyze. There also is an increasing reuse of scientific data, so people are using each other's data for purposes that were not necessarily originally intended.

The data publishing in this exponential world is also changing very dramatically. The big astronomy projects typically are undertaken by collaborations of 60 to 100 people, who work for 5 or 6 years to build an instrument that collects the data and who then operate it for at least that long, because otherwise it would not be worth investing that much of their time. Once they have the data, they keep using them and eventually publish the data and their analyses. They organize the data in a database and make them accessible on their Web sites. When the project ends, the scientists go on to other projects, and at that point they are ready to hand over the data to some big national archive or centralized data storage facility. After that, the scientists continue to use the archived data.

Why are the roles changing? The exponential growth makes a fundamental difference. There is also more responsibility placed on the research projects. Astronomers and other scientists are learning how to become publishers and curators, because they do not have a choice if they want to make

---

[3]This section is based on the presentation by Alex Szalay, Alumni Centennial Professor, Department of Physics and Astronomy, The Johns Hopkins University.

their data public. More standards and more templates would help with this.

There also is a major trend toward making high-capacity computing more distributed. This is called grid computing, where the computing is distributed all across the Internet at multiple sites, and people can borrow time on central processing units (CPUs) whenever they need it. The people who talk about grid computing, however, tend to think only about harvesting the CPUs; they do not think about the hundreds of terabytes or possibly petabytes of data behind it, because we currently lack the bandwidth and cannot move the data to the computers.

Alex Szalay, of the Johns Hopkins University, and Jim Gray, of Microsoft Research, have begun a project to make these astronomical data understandable and useable by high school students. They opened their Web site in 2001, and after 2 years they have about 12 million pages online and get about 1 million hits per month. The site is used by high school students who are learning astronomy, but who are also learning the process of scientific discovery, using up-to-date data—data that are as good as any astronomer can get today.

Astronomical observations are diverse and distributed, with many different instruments constantly observing the sky from all the continents, in different wavelengths, and producing a lot of data. This all adds up to the concept of a "virtual observatory." The vision for the virtual observatory was to make the data integration easy by creating some standard interfaces and to federate the diverse databases without having to rebuild everything from scratch. Astronomers also wanted to provide templates for others, for the next generation of sky surveys, so they could build it the right way from the beginning.

This idea has taken off. The National Science Foundation funded a project for building the framework for the national virtual observatory, which involves all the major astronomy data resources in the United States—astronomy data centers, national observatories, supercomputer centers, universities, and people from various disciplines, including statistics and computer science. There is now a formal international collaboration— the International Virtual Observatory Alliance.

Publishing this much data requires a new model. It is not clear what this new model is, however, so astronomers are trying to learn as they go. There are multiple challenges in the use of the data for different communities. There are data mining and visualization challenges—how to visualize such a large, distributed complex set of databases. There are important educational aspects of it; students now have the ability to see the same data as professional astronomers do. And there is very much more data coming, petabytes per year by 2010.

Indeed, the same thing is happening in all of science. Science is driven by Moore's law, whether in high-energy physics, genomics, cancer research, medical imaging, oceanography, or remote sensing. This also shows that there is a new, emerging kind of science. We are now generating so much data, both real data and in simulations, that we need a combination of theory, empirical computational tools, and also information management tools to support the progress of science.

GENOMIC DATA CURATION AND INTEGRATION WITH THE LITERATURE[4]

One of the driving forces for most scientists, certainly those in biological research, is that science is becoming more data intensive. This means that researchers are generating more data for each paper, but they are also using more data from others for each paper. That has an impact on both the factual databases and the literature. In order to make this work, we will need to have deeper links and better integration between the literature and the factual databases to improve retrieval from both, and to improve their actual usability and the extraction of value from them. The quality of the factual data can be very much improved if one can get a tighter integration between the literature and the databases.

In most areas of biology, as in all other areas of science, the increase in the amount of data is exponential. For example, the number of users per weekday at the National Center for Biotechnology Information (NCBI)[5] Web site is more than 330,000 with different IP addresses, and this is growing. They are using the NCBI data to design experiments and write their articles. The electronic journals now have many links or the number of identifiers from databases that authors are including in their papers. This is also the case with supplementary data files.

At PubMed Central,[6] which is NIH's archive for the biomedical literature, there are many links and other functions as well. One can, for ex-

---

[4]This section is based on a presentation by David Lipman, director of the National Center for Biotechnology Information.

[5]See http://www.ncbi.nlm.nih.gov/ for additional information about the NCBI.

[6]For additional information about PubMed Central, see http://www.pubmedcentral.nih.gov/.

ample, link from a full-text article to get all of the referenced articles that are in PubMed from it, as well as a variety of other databases.

By having this fairly fine level of integration and links between the literature and factual databases, the article has a higher value, because not only can the reader understand better the point that the author was trying to make, but can go beyond that point and look at newer information that is available.

ISSUES RAISED IN THE DISCUSSION

The Need for Federal Coordination and Investment in the Cyberinfrastructure

The exponential data growth in many fields illustrates that the challenges and opportunities include going to higher performance networks, higher speed computers, and greater capacity storage, but to do that together with functional completeness by having the complete range of services. There is exciting potential for multiuse technologies: The common underlying infrastructure is serving the leading edge of science, and making the learning of science more vivid, more authentic, and exciting. Although a major investment is needed to create this infrastructure, once it is created, as the astronomy example illustrates, leading-edge teams or individual amateurs can make seminal and important contributions to science, provided they are given open access to these data and to the tools.

Both the opportunities and the challenges illustrate the urgency for some leadership in this area, however. If we do not get the right investments or the right synergy among domain scientists, librarians, information specialists, and technologists, we could end up with sub-optimal solutions or solutions that do not scale. Worst of all, we can end up with Balkanized environments that cannot interoperate and that result in enormous lost opportunity costs.

Quality Control and Review of Data in Very Large or Complex Databases

In astronomy, the condition for putting contributed data online is that they are provided with adequate metadata in certain formats. This keeps a lot of people out whose data are not of high enough quality, and who have not documented them sufficiently.

At the NCBI, there is a comprehensive set of curated sequences. For these molecular biology data, there are two related sets, the archived set, which represents what scientists provided at the time, and the curated set, which contains what is supposed to be the best version of the data.

An important point relates to the difference between astronomy and biology. Astronomy has an organizing principle, in that there are space-time coordinates, which is largely agreed upon. The biology data do not have that. One of the difficulties of a project like STKE and any of the other projects in functional genomics is that it is much more difficult to use cross-validation to fully assess the quality of the data. Because there are high-throughput methods in biology that are at the level of function, it is a challenge to deal adequately with quality.

Data-Mining Restrictions from Proprietary STM Information

How might we automatically download, create, and centralize a repository of identified articles if the intent is to extract data and republish subsequently extracted facts?

There are restrictions on the access to and the use of the data from proprietary sources. How can data mining to extract related facts, which presumably could have important results, be done when these information sources are still subject to the proprietary model?

One answer might be that perhaps this tension is good. It forces publishers to think about what they are doing and the basic goals they have. If their purpose is to help researchers be more efficient, and to advance science and serve society, then the new technologies and new business models need to be examined much more thoroughly.

Publishing Large and Complex Data Sets for Knowledge Discovery

The methods for organizing and labeling the huge data sets reflect current knowledge in the field. One question that arises is whether the availability of all these wonderful data to so many people enables researchers to make quantum leaps in their knowledge of phenomena. Or, is there a risk that because the data are organized according to what we understand now, it might be tempting just to stay in that vineyard and not make big advances or face big changes in modes of thinking? How might this issue be addressed, particularly for disciplines that are not as far along as those that are putting together these huge data sets?

In the case of the Sloan Digital Sky Survey, which is now about 40 percent complete, after 2 nights of operation astronomers found 6 out of 10 of the most distant quasars in the universe. This is a small telescope by astronomy standards, but it shows that astronomers are efficient in finding new objects by organizing the data and scanning through the outliers.

For databases like GenBank, the U.S. sequence database, and for most other databases in the life sciences, there are multiple sites. GenBank collaborates with similar centers in Japan and the United Kingdom. The data are exchanged every night, but they have different retrieval systems and different ways of organizing access into them. Furthermore, people can download large subsets or the entire database and do whatever they want, including making commercial products.

With the STM literature, if we had more open-access journals, and there were multiple sites that provided comprehensive access, one would see different capabilities. PubMed Central is working with a group in France to set up a mirror archive that would organize that data in a different retrieval system. Open access to the data allows for multiple archives that are comprehensive and provides different ways to get at that information.

The questions posed above raise a deeper issue, however. In biology, gene ontology is a way to make it easier for people to see what is there, and to move the information around and to understand it. This represents a trade-off between what we understand now and the kind of astronomical discoveries referred to above.

Right now, there is a huge amount of interest in ontologies in biology, and some of it may be misplaced. One of the reasons researchers focused on molecular biology was that they really did not understand enough from the top down. If one looks at proteins or genes that are involved in cancer, you find those that are a part of glycolytic pathways, and so forth. It is not clear how much these ontologies assist in finding and understanding things, and how much they obscure new discoveries. As long as we maintain access to the data in a very open way, and people can download and do almost whatever they want with them, then if they want to ignore something like an ontology, they can do that.

Another issue is the migration away from text surrogates to full texts that allow computation. That capability is having a radical effect in many fields. It is useful to be able to find things by doing computation and searching on the full text of scholarship, as opposed to being locked into some kind of a classification structure for subject description that some particular group used in creating surrogates. That is a theme heard from scholars in every field from history all the way through biology. It is really quite a striking example of how things are changing.

Transformation of Archiving in the Knowledge Discovery Processes

If we are moving from publication as product to publication as process, should we make a similar transformation in archiving? Or, should we still want to archive products that may be extracted from that process? An example in STKE would be the "Viewpoints" from *Science Online*, which are snapshots in time. Archiving the best contributions could be useful, and these then might even be published in print form. So far, however, the focus for the STKE Viewpoints is not archival; although it does serve that purpose, it is more to give the authorities some recognition.

There is an enormous stream of digital objects that could be created by knowledge discovery processes that are mediated through technology. One may want to archive not only these objects in temporal streams, but the relationships between the objects and the social context in which they are created.

One of the most profound ideas about this came from John Seeley Brown, who posited that perhaps the most important aspect of this technology-mediated way of work is not just relaxing the constraints of distance and time, enhancing access, and so forth, but the possibility of actually archiving the process itself, not just sampling the artifacts along the way.

In areas of ubiquitous computing, people could subsequently return and actually mine processes and extract new knowledge that otherwise has been left on the table. It is an extension of the notion of data mining into knowledge process mining, so it can get very abstract, but we can start to see that it is not just fanciful but something to think about. People who are interested in long-term preservation need to consider huge repositories that take into account not only the basic information but the social processes by which the information is disseminated, manipulated, and used.

Increasing Data and Lost Opportunities

According to statistics presented by Donald King, over the past 15 years, scientists have increased the amount of time that they devote to their work by about 200 hours a year. Scientists therefore are reaching the limits of their capacity for how much time they can spend on their work. Most of those additional 200 hours are devoted to communicating. The number of scientists only increases about 3 percent a year, however, which means that the population of scientists doubles about every 15-20 years or so, yet some of the information we are gathering doubles every year. The limitations of the capacity of the human intellect to work with these data may therefore be a concern. The scientific community may wish to increase the number of scientists who work with the data and the information infrastructure. It seems that there must be lost opportunities.

At the same time, of course, scientists do adapt to dealing with large data sets. If presented with more computing power and data, scientists ask different kinds of questions. It may take a long time, however, before more scientists within the community shift and start to think of different kinds of questions. A few pioneers start to think a new way, and then it starts to take hold.

However, because of this avalanche of data, if all the data get properly published, it will cause another fundamental sociological change. Today in science, many people train a lifetime to build up their own suite of tools that they apply to the raw data that they can get. Through these new digital resources, if one can get reasonably polished data, one can think more about the discovery itself; a researcher does not have to spend so much time with the mechanics of scrubbing the raw data and converting them into useable form. People will be much less reluctant to cross disciplinary boundaries if the data are all available in a ratified and documented form.

There also is a difference between organizing data sets and making them useable to other people, which is a challenge, and finding and extracting what one thinks about that data set and getting the results published. The article represents what the scientist did; it conveys the knowledge. That can only be done so fast.

The Role of Journals in the Successful Development of Databases in Molecular Biology

The enormous success of factual databases in molecular biology and the scientific community's reliance upon them are largely results of the collaborative effort the journals have made with regard to their requirement that the data be deposited as a prerequisite to publication. In the area of protein structure, for example, the data that crystallographers gathered were held and not deposited. The pressure from peers to a certain extent, but more important, from journals, resulted in those data being deposited as a condition of publication, which makes the databases in this area as robust as they are.

Despite the fact that databases are useful, scientists often do not want to spend their time on that. What they are judged by is not what they put into a database, but what they publish. The *Journal of Biological Chemistry*, a not-for-profit journal published by the American Society for Biochemistry and Molecular Biology, was one of the pioneer journals in requiring submission to the sequence databases and in getting essentially 100 percent compliance.

7

Symposium Wrap-Up

MODERATOR'S OVERVIEW[1]

The changes in the publishing system that are being driven by online journals were discussed extensively throughout the symposium. There is a range of issues that requires close attention, some of which are summarized here. For instance, it is clear that the costs of online-only journals are less than print plus online. Nevertheless, print versions seem unlikely to be eliminated completely across all of STM publishing in the near future, and so the costs of a dual system persist for most publishers.

There are economies of scale that tend to favor publishers of large numbers of STM journals, and these economies are not available to small and mid-sized publishers. As a result, some cooperation and grouping of content have arisen within some sectors in order to mimic or replicate these economies. Recent mergers and acquisitions within the STM journal market have heightened smaller publishers' concerns about their longer term viability as independent entities. Most of the broad range of online business models are still quite experimental and seem likely to diversify and hybridize over time as the publishing system develops and new advances emerge.

---

[1]The Moderator's Overview was given by Mary Waltham, publishing consultant, and member of the steering committee for this symposium.

With regard to archiving, questions remain about who will be responsible for the long-term preservation of journal content—publishers, national libraries, or other third parties? What will be archived—all of the journal content or just the research content? Who will curate the journal archive and ensure it migrates to appropriate platforms as technology evolves? Who will pay?

Filtering and quality control of the information were identified as being central to the publishing process. In the future, with more open access to all types of information and data, who will provide reliable and consistent filtration and quality control, and who will pay for that?

Increased online access results in increased usage of information. Speakers talked about the likelihood of publishing becoming a more disaggregated process with separate pieces of the continuum done by different groups—from content creation through dissemination. The journal as a package of information is not granular enough and so further unbundling of information is taking place. Customers and users want more granularity in the online environment than a journal issue represents.

There also was discussion of who needs copyright as opposed to "wants it" or "uses it." Copyright is very dependent on both the author and the mission of the publisher with whom they may be working.

Finally, the continuous online publishing process means that documents may no longer be static but evolve through time because addition, annotation, and revision are simple. Interoperability and common standards are essential to bring together and integrate information and to provide a dynamic reference tool. Achieving integration is essential for making the optimal use of scientific information.

A key point for publishers is, where do they add value to the publishing process, and is that value added where users and customers want it? The role of publishers must continue to change to meet the needs of the research community.

The remainder of this Wrap-Up session highlighted a number of other key issues raised during the meeting from the perspective of a working scientist, a university administrator, and a librarian.

REFLECTIONS OF A PRACTICING SCIENTIST[2]

Within the physics community at Stanford University online journal access has already replaced print. Open access to scientific information appears to be the next inevitable step. It is what the academic community is now looking for and will finally insist on, although getting to that point will not be simple. This transition will generate tension, which will require strong and well-informed leadership to ensure that ultimately that tension is creative for science and those involved with it. STM journal publishing will not go away; rather, the current changes in the system are forcing a closer examination of where value is added in the process and what the alternatives are. New modes of value added develop rapidly online, and some of these will persist.

The efficiency of information transfer is enhanced by online formats—it is faster and there is more of it—but this is not an ultimate good without increased understanding as distinct from knowledge and acquisition of facts. Although the current online tools enable rapid access, they do not by themselves improve or accelerate the ability to form good judgments. Students, in particular, struggle with trying to understand research materials. Overall, however, the implications of electronic publishing have not received sufficient attention in the scientific and university communities.

PERSPECTIVES OF A UNIVERSITY ADMINISTRATOR[3]

Discussions about the future of journal publishing and its models have two crucial elements. The first element of a new system is that it will drive costs out of the old system, although electronic publication of course is not cost free. Recovering those costs and sharing them in an equitable and sustainable manner are separate issues, however, and the second key element.

The open-access movement tends to shift costs away from the user and toward the producer of information. The experience in subscription-based journal publishing over the past 15-20 years has demonstrated that among the features of high-priced journals is a form of quality control. When university libraries cut their budgets for serials acquisition and reduce the

---

[2]Malcolm Beasley, Theodore and Sydney Rosenberg Professor of Applied Physics, Stanford University.

[3]James O'Donnell, provost, Georgetown University, and cofounder in 1990 of the Bryn Mawr Classical Review, an online book review journal for classical studies.

number of titles they acquire, they are exercising a form of quality control over what they will accept. If the "big deal" is fraying around the edges, that will transmit itself back to the publishers as news about which journals are to be sustained and which journals are not. The closing of print journals and the cutting back of titles by publishers over the past two decades have been achieved mainly as a result of publishers' mergers and acquisitions and the recognition that not every title is sustainable.

Does the open-access model deliver superior quality over the commercial model? As one argues about business models and recovery of costs, it remains an open question whether a change in the system will improve or degrade the quality of scholarly communication. As long as that issue remains open, there is no forcing argument to use in favor of one model or the other. It is important, therefore, to continue to assess what the effects on quality of information, timeliness of access, and quality of peer review in the different models are.

There is an emerging differentiation of the products that scholars, scientists, and publishers want to have supported. At least three different kinds of information were discussed at this symposium. At one end, there is the timeliest of information services, providing the linked news from the research front as rapidly as possible with little consideration for archiving or price, but simply getting the most recent results disseminated as fast as possible so that science can progress. At the other end of the life expectancy of that information is when it has become an artifact, something to be preserved, maintained, and sustained long after its commercial life, or perhaps even after its scientific life, has been exhausted.

That first information service tends to be market based. That last artifact service is not market based at all, but is something done out of noblesse oblige for the greater good of the community. Between them there is a borderline area, where the information service itself needs to be mediated to those who have limited access to the market. An effective scientific, technical, and medical information system will address inequities in the market and find places at the information table for those who do not have the market clout behind them. Understanding that differentiation of product, which is an increasing differentiation in the electronic environment, increasing not only among products but increasingly differentiated among disciplines, will be an important part of understanding what a new system of information can be like. It is time to begin disaggregating the scholarly and scientific publishing crisis, if there is one, into problems that can be addressed in rational and coherent kinds of ways.

A RESEARCH LIBRARY VIEWPOINT[4]

The progress of science requires access both to raw material and to evaluated judgments and conclusions. It appears that the raw material and the evaluated conclusions coming out of scholarly research are bifurcating into two fairly distinct realms. One of them is a quite tightly controlled, often highly priced, peer-reviewed literature and the other is the minimally controlled, scholar-managed, open-access publication and database regime. Both environments present challenges to the universities. One could generalize from the symposium that some publishers are making every effort to increase control over the content that they publish and to expand their reach both in time and in format.

The university perspective on the value chain of the process by which new information is created is very different from the value chain perspective that one hears either from authors or from publishers. From the university perspective, the university builds the infrastructure and provides an opportunity for faculty to develop curricula and conduct research. The universities attract and admit the students, especially the graduate students who also conduct some of the research and represent the next generation of the research community. The universities provide access to information at a cost that seems to them to be reasonable, as a percentage of their overall annual operating budget, and they delegate that responsibility to their research libraries.

Not everyone in research libraries believes that the subscription model as a way of acquiring information is fundamentally broken. However, the costs to the libraries under the subscription model are in many cases considered excessive, and the terms and conditions under which information is allowed to be disseminated and used on university campuses are onerous because licensing agreements control which subsets of a community can actually look at the information. It is the combination of the price and of the constraints on the information that represent problems in the system from the library perspective.

Over the past 10 or 15 years publishers have developed independent business models, which do not have a symbiotic relationship with the university. It is in that environment that the costs have gone up and that the terms and conditions of use have changed. There is now a very clear bound-

---

[4]Ann Wolpert, director of libraries, Massachusetts Institute of Technology.

ary between the universities and the publishers that capture the work that comes out of universities, evaluate the quality of that work, and publish it.

The conundrum for universities has several dimensions. For many publications, the costs are simply too high for the value received, and the licensing conditions are problematic, in terms of what we can do, particularly with digital information when it is delivered to our campuses.

The intellectual property environment is not only incomprehensible to the average faculty member and student, but it places what happens at universities at risk. Any legal regime that varies by circumstance is not a particularly useful one for the users. It seems clear that intellectual property law that is designed to meet the needs of the entertainment industry and the international publishing conglomerates is not particularly conducive to facilitating the needs of the academic community.

It also is apparent that the functions of scholarly communication and publication are diverging. Traditionally, communication with colleagues was through publications; now it is quite clear that researchers, professors, and students can communicate outside of the formal publication record. The formal publication record is moving in many cases off to one side, which again affects the question of how much value we should put in the formal record of advances in a discipline, if in fact most of the communication is happening some other way.

It is difficult to determine what reasonable standards and norms might be for the cost of peer-reviewed publication. Those costs vary tremendously, and we do not know what drives them. We also do not know what drives the cost of print as opposed to the electronic publication. So it is very hard for us to think logically without the kinds of norms and standards that one can get from most other industries.

Finally, we do not know what new models of peer review and recognition might be developed for open-source publications, which is an area that requires more attention.

Appendixes

A

Symposium Agenda

SYMPOSIUM ON ELECTRONIC SCIENTIFIC, TECHNICAL,
AND MEDICAL JOURNAL PUBLISHING AND ITS
IMPLICATIONS

Main Auditorium
The National Academies
2100 C Street, NW
Washington, DC 20418

Monday, May 19

8:00   *Registration and Continental Breakfast*

8:30   Welcoming Remarks
       Bruce Alberts, president, National Academy of Sciences

8:45   Symposium Overview
       *Edward Shortliffe, professor and chair, Department of Medical
       Informatics; deputy vice president for Information Technology
       Health Sciences Division, Columbia University; and symposium chair*

9:00   Keynote Address
       *James J. Duderstadt, president emeritus and University Professor of
       Science and Engineering Millennium Project, University of
       Michigan*

Panel 1: Costs of Publication
Moderator: Floyd Bloom, The Scripps Research Institute

9:30   Opening Remarks by Moderator

9:35   Overview Presentation
        Michael A. Keller, CEO, HighWire Press

9:55   Comments by Panel Participants
        Kent R. Anderson, publishing director, *New England Journal of Medicine*
        Robert Bovenschulte, director, Publications Division, American Chemical Society
        Bernard Rous, deputy director/electronic publisher, Association for Computing Machinery
        Gordon Tibbitts, president, Blackwell Publishing USA

10:25   *Break*

10:45   Discussion of Issues

12:00   *Lunch*

Panel 2: Publication Business Models and Revenue
Moderator: Jeffrey MacKie-Mason, University of Michigan

1:00   Opening Remarks by Moderator

1:05   Comments by Panel Participants
        Joseph J. Esposito, president and CEO, SRI Consulting
        Wendy Pradt Lougee, director, University of Minnesota Library
        Brian Crawford, vice president and general manager, Life and Medical Sciences, John Wiley & Sons
        Patrick O. Brown, professor of biochemistry, Stanford University

1:55   Discussion of Issues

3:10   *Break*

Panel 3: Legal Issues in Production, Dissemination, and Use
Moderators: Ann Okerson, Yale University, and Jane Ginsburg,
Columbia Law School

3:30   Copyright Basics: Ownership and Rights
        Jane Ginsburg, Morton L. Janklow Professor of Literary and
        Artistic Property Law, Columbia Law School

3:50   Economic and Non-Economic Rewards to Authors
        Michael Jensen, Jesse Isidor Straus Professor of Business
        Administration, Emeritus, Harvard Business School

4:10   Licensing
        Ann Okerson, associate university librarian for collections and
        technical services, Yale University

4:30   Discussion of Issues

5:45   *Adjourn*

6:00   *Reception, National Academy of Sciences' Great Hall*

Tuesday, May 20

8:00   *Continental Breakfast*

Panel 4: What Is Publishing in the Future?
Moderator: Daniel E. Atkins, University of Michigan

8:30   Opening Remarks by Moderator

8:35   Institutional Repositories
        Hal Abelson, Class of 1922 Professor of Computer Science and
        Engineering, Massachusetts Institute of Technology

8:50   Preprint Servers and Extensions to Other Fields
        Richard E. Luce, research library director, Los Alamos National
        Laboratory

9:05 Implications of Emerging Recommender and Reputation Systems
Paul Resnick, associate professor, University of Michigan

9:20 Discussion of Issues

10:35 *Break*

Panel 5: What Constitutes a Publication in the Digital Environment?
Moderator: Clifford Lynch, Coalition for Networked Information

10:55 Opening Remarks by Moderator

11:00 Signal Transduction Knowledge Environment
Monica Bradford, executive editor, Science

11:15 Publishing Large Data Sets in Astronomy—
The Virtual Observatory
Alex Szalay, Alumni Centennial Professor, Department of Physics and Astronomy, The Johns Hopkins University

11:30 Data Curation and Integration with the Literature
David Lipman, director, National Institutes of Health/National Center for Biotechnology Information

11:45 Discussion of Issues

1:00 *Lunch*

Panel 6: Wrap-Up Session
Moderator: Mary Waltham, publishing consultant

1:55 Opening Remarks by Moderator

2:00   Symposium Summaries
        Malcolm R. Beasley, Theodore and Sydney Rosenberg Professor
           of Applied Physics, Stanford University
        James J. O'Donnell, provost, Georgetown University
        Ann Wolpert, director of libraries, Massachusetts Institute of
           Technology

2:30   Discussion of Issues

3:10   Closing Remarks by Symposium Chair, *Edward Shortliffe*

3:15   *Adjourn*

B

Biographical Information for Speakers and Steering Committee Members

Hal Abelson is professor of electrical engineering and computer science at Massachusetts Institute of Technology (MIT) and a fellow of the Institute of Electrical and Electronics Engineers (IEEE). He is winner of several teaching awards, including the IEEE's Booth Education Award, cited for his contributions to the teaching of undergraduate computer science. Professor Abelson's research at the MIT Artificial Intelligence Laboratory focuses on "amorphous computing," an effort to create programming technologies that can harness the power of the new computing substrates emerging from advances in microfabrication and molecular biology. He is also engaged in the interaction of law, policy, and technology as they relate to societal tensions sparked by the growth of the Internet, and he is active in projects at MIT and elsewhere to help bolster our intellectual commons. Professor Abelson is a founding director of the Free Software Foundation and a founding director of Creative Commons. He also serves as consultant to Hewlett-Packard Laboratories. At MIT, Professor Abelson is codirector of the MIT-Microsoft Research Alliance in educational technology and cohead of MIT's Council on Educational Technology.

Bruce Alberts, president of the National Academy of Sciences, is known for his work both in biochemistry and molecular biology, in particular for his extensive study of the protein complexes that allow chromosomes to be replicated. Dr. Alberts graduated from Harvard College and earned a doctorate from Harvard University in 1965. He joined the faculty of Princeton University in 1966 and after 10 years moved to the Department of Biochemistry and Biophysics at the University of California, San Francisco, where he became chair. He is one of the original authors of *The Molecular Biology of the Cell*, through four editions the leading advanced textbook in this important field. His most recent text, *Essential Cell Biology* (1998), is intended to present this subject matter to a wider audience. Dr. Alberts has long been committed to the improvement of science education, dedicating much of his time to educational projects such as City Science, a program that seeks to improve science teaching in San Francisco elementary schools.

Kent R. Anderson is the publishing director for the *New England Journal of Medicine*. Prior to joining the journal, he was director of medical journals at the American Academy of Pediatrics. He has been in health care publishing for more than 15 years and has worked as a writer, editor, designer, production manager, copy editor, managing editor, and publisher. He has also worked in continuing medical education, launched a half-dozen successful new titles, and contributed to numerous online publishing initiatives.

Daniel E. Atkins* is a professor in the School of Information and in the Department of Electrical and Computer Engineering at the University of Michigan (UM), Ann Arbor. He began his research career in the area of computer architecture and did pioneering work in high-speed computer arithmetic and parallel computer architecture. He has served as dean of the College of Engineering and more recently as the founding dean of the School of Information at UM. He is now director of the Alliance for Community Technology (ACT), an international partnership with philanthropy for research and development in the use of information and communication technology (ICT) to further the mission of educational and other nonprofit organizations. Dr. Atkins does research and teaching in the area of distributed knowledge systems. He has directed several large experimental digital library projects as well as projects to explore the application of "collaboratories" to scientific research. He has recently served as chair of the National Science Foundation Advisory Panel on Cyberinfrastructure. The panel issued a report in February 2003 recommending a major Advanced Cyberinfrastructure Initiative intended to revolutionize science and engineering research and education. He also serves regularly on panels of the

---

* Indicates member of the Symposium Steering Committee.

National Academies exploring issues such as scholarship in the digital age, the future of scholarly communication, and the impact of information technology on the future of higher education. He is coauthor of a recent book entitled *Higher Education in the Digital Age: Technology Issues and Strategies for American Colleges and Universities.* He serves as a consultant to industry, foundations, educational institutions, and government.

Malcolm R. Beasley is professor of applied physics in the Geballe Laboratory for Advanced Materials at Stanford University. He received his B.Eng. and his Ph.D. in physics from Cornell University. He then went to Harvard University as a research fellow and subsequently became a member of the faculty. In 1974 he joined the faculty of Stanford University where he became a full professor of applied physics in 1979. He served as the chairman of the Department of Applied Physics at Stanford from 1985-1989. In 1990 he was named the Theodore and Sydney Rosenberg Professor of Applied Physics. From 1992 to 1998 he served as director of the Center for Materials Research. And from 1998 to 2001 he served as dean of the School of Humanities and Sciences. Professor Beasley is a member of Tau Beta Pi, the IEEE, the National Academy of Sciences, and the American Academy of Arts and Sciences, and a fellow of the American Physical Society and the American Association for the Advancement of Science. He is the recipient of the Dean's Award for Superior Teaching at Stanford University. He has served as a consultant to the National Science Foundation, Department of Energy, Defense Advanced Research Projects Agency, and various industrial laboratories. He has also served on various panels of the National Research Council of the National Academies. He was an elected member of the Board of Trustees of Associated Universities, Inc., for the period 2003-2005. Professor Beasley's research interests are in materials physics with an emphasis on basic and applied superconductivity, in particular high-temperature superconducting materials and applications, and the development and application of advanced thin-film deposition techniques for complex materials.

Floyd Bloom* is chairman of the Department of Neuropharmacology at the Scripps Research Institute. He previously served as director of Behavioral Neurobiology at the Salk Institute and as chief of the Laboratory of Neuropharmacology of the National Institute of Mental Health. Dr. Bloom is a member of the National Academy of Sciences and the Institute of Medicine (IOM). He has received many awards, including the Pasarow Award in neuropsychiatry and the Hermann von Helmholtz Award, as well as a number of honorary degrees from major universities. He served as editor-in-chief of *Science* magazine from 1995-2001 and currently is the president of the American Association for the Advancement of Science Board of Directors.

Robert Bovenschulte is director of the Publications Division of the American Chemical Society (ACS), which publishes journals, magazines, books, and electronic products. Prior to joining ACS in 1997, he was vice president for publishing at the Massachusetts Medical Society, owner of the *New England Journal of Medicine* and other medical publications. His career spans scholarly, professional, trade, college, and school segments of the industry. He is a frequent speaker and moderator at conferences of publishers and librarians. Mr. Bovenschulte has served as chair of the Executive Board of the International Association of Scientific, Technical, and Medical Publishers, chair of the Board of Directors of the Copyright Clearance Center, chair of the Executive Council of the Association of American Publishers' Professional and Scholarly Publishing Division, and member of the Board of Directors of the Council on Library and Information Resources.

Monica Bradford is the executive editor of the international journal *Science* (published by the American Association for the Advancement of Science). In this position she oversees the peer-review and selection of manuscripts; the copyediting and proofreading process; and the design, production, and manufacture of the print product. Over the past few years, Ms. Bradford has been heavily involved in the development of *Science* Online. In particular, she has helped create a new line of digital products referred to as online knowledge environments. *Science*'s STKE and SAGE KE, the first two products in this line, are directed at research scientists. In addition, Ms. Bradford administers the APBiotech & *Science* Prize for Young Scientists in Molecular Biology. Prior to joining the staff of *Science* in 1989, Ms. Bradford worked for the Publications Division of the American Chemical Society for 9 years. She holds a bachelor's degree in chemistry from St. Mary's College, Notre Dame, IN, and has done graduate work in management at the University of Maryland. Ms. Bradford was a member of the Board of Directors of the Council of Biology Editors, served as vice-chair of the Scientific Publishing Board of the American Heart Association, and is a member of the Society for Scholarly Publishing.

Patrick O. Brown is a professor of biochemistry at Stanford University School of Medicine and a Howard Hughes Medical Institute Investigator at the Stanford Medical School. He received his B.A. in chemistry, and M.D. and Ph.D. in biochemistry from the University of Chicago. He did his graduate work with Nick Cozzarelli, studying the basic molecular mechanisms of DNA topoisomerases. Following a residency in pediatrics at Children's Memorial Hospital in Chicago, he began studies of the mechanism of retroviral integration as a postdoctoral fellow at the University of California, San Francisco, working with J. Michael Bishop and Harold Varmus. Dr. Brown is a member of the National Academy of Sciences. Dr. Brown's current research uses DNA microarrays and other "genomic" approaches to explore a wide range of fundamental questions in gene regulation, cell biology, physiology, development, and medicine. For the past several years he has been working to promote open, unrestricted access to scientific and scholarly publications. He is a cofounder and codirector of the Public Library of Science, a nonprofit, open-access scientific publisher.

Brian Crawford is vice president and publishing director for Global Life and Medical Sciences within the Scientific, Technical, and Medical (STM) publishing operations of John Wiley & Sons, Inc. He holds overall responsibility for coordinating the strategic development and ongoing editorial management of Wiley's international Life and Medical Sciences publishing. Experienced in subscription journal, book, and new media publishing, Dr. Crawford has spent nearly two decades within the commercial STM information industry, most recently having served in several management roles at Wiley. Before joining Wiley, he held the position of vice president and editor with Alan R. Liss, Inc. (1988-1989), a privately held book and journal publishing firm that merged to become the wholly owned Wiley-Liss, Inc. subsidiary. Mr. Crawford began his publishing career in 1985 as an acquisitions editor within the Journals Publishing Division of Academic Press, Inc. (at that time a part of Harcourt Brace Jovanovich; now a part of Reed Elsevier). Prior to entering scientific publishing, Dr. Crawford was active in both scientific research and teaching. He had the distinction of being the first biologist to be appointed as a J. Robert Oppenheimer Fellow of the Los Alamos National Laboratory, where he was a member of the scientific research staff within the Genetics Group of the Life Sciences Division from 1981-1984, and helped to launch the Department of Energy (DOE)-sponsored human genome project. He received his Ph.D. in the biochemical and biophysical sciences at the Johns Hopkins University School of Public Health, specializing in cellular and molecular aspects of cancer genetics, under sponsorship from the National Cancer Institute. He received his B.S. cum laude in chemistry from the University of Maryland at College Park. Dr. Crawford is a member of the Board of Directors of the American Medical Publishers Association (AMPA; president-elect in 2002-2003) and the Executive Council of the Professional and Scholarly Publishing Division of the Association of American Publishers (PSP/AAP; 2001-present).

James J. Duderstadt is president emeritus and University Professor of Science and Engineering at the University of Michigan. He received his B.Eng. in electrical engineering from Yale University and his Ph.D. in engineering science and physics from the California Institute of Technology. Dr. Duderstadt joined the faculty of the University of Michigan in 1968 as professor of nuclear engineering. He became dean of the College of Engineering in 1981 and provost and vice president for academic affairs in 1986. He was appointed president of the university in 1988 and served in this role until July 1996. He currently holds a university-wide faculty appointment. Dr. Duderstadt's teaching and research interests have spanned a wide range of subjects in science, mathematics, and engineering, including work in areas such as nuclear energy, lasers, computers, science policy, and higher education. During his career, Dr. Duderstadt has received numerous awards for his research, teaching, and service activities, including the National Medal of Technology for exemplary service to the nation. Dr. Duderstadt has served on and chaired numerous public and private boards, including the National Science Board and the National Academy of Engineering. He also serves as a director of Unisys and CMS Energy. He currently chairs several major national study commissions in areas including federal research policy, higher education, information technology, and nuclear energy.

Joseph J. Esposito is president and CEO of SRI Consulting, the leading publisher of syndicated research for the global chemical industry. Over the course of his career, Mr. Esposito has been associated with various publishers in all segments of the industry and was involved from an early time with new media publishing. He has served as an executive at Simon & Schuster and Random House, as president of Merriam-Webster, and CEO of *Encyclopaedia Britannica*, where he was responsible for the launch of the first Internet service of its kind. Mr. Esposito has also worked extensively in the technology industry as a consultant, with such clients as Microsoft and Hewlett-Packard, and formerly ran the Internet communications company Tribal Voice. His primary area of concentration is the development of strategy and business models for the dissemination of digital content. He has participated in numerous trade shows and has written extensively in trade magazines and journals (see, for example, the essay "The Processed Book" in the March 2003 issue of *FirstMonday* at www.firstmonday.org). Mr. Esposito is currently researching new economic models for a post-copyright age. He can be reached at jesposito@sric.sri.com.

Jane Ginsburg* is Morton L. Janklow Professor of Literary and Artistic Property Law at Columbia Law School. Ms. Ginsburg received her J.D. from Harvard Law School and her Doctor of Law from the Universite de Paris II. She served as clerk to Judge John J. Gibbons, U.S. Court of Appeals for the Third Circuit and spent 3 years in private practice before turning to teaching. Her principal areas of interest and expertise are in intellectual property, comparative law, private international law, and legal methods. She has published extensively on copyrights and intellectual property and serves on the editorial boards of several intellectual property journals in the United States and abroad.

Michael C. Jensen is the managing director of the Organizational Strategy Practice at the Monitor Company and Jesse Isidor Straus Professor of Business Administration emeritus of the Harvard Business School. Professor Jensen joined the faculty of the Harvard Business School in 1985. In 1999, he left Harvard to assume his current position at the Monitor Company. He was LaClare Professor of Finance and Business Administration at the William E. Simon Graduate School of Business Administration, University of Rochester, from 1984-1988, professor from 1979-1984, associate professor from 1971-1979, and assistant professor from 1967-1971. He founded the Managerial Economics Research Center at the University of Rochester in 1977 and served as its director until 1988. Professor Jensen earned his Ph.D. in economics, finance, and accounting, and his M.B.A. in finance from the University of Chicago, and an A.B. degree from Macalester College. He also has been awarded several honorary degrees and served as a Visiting Scholar at the Tuck School of Business at Dartmouth College from July 2001 to June 2002. Professor Jensen is the author of more than 50 scientific papers, in addition to numerous articles, comments, and editorials published in the popular media on a wide range of economic, finance, and business-related topics. He is author of *Foundations of Organizational Strategy* (Harvard University Press, 1998) and *Theory of the Firm: Governance, Residual Claims, and Organizational Forms* (Harvard University Press,

2000). He is editor of *The Modern Theory of Corporate Finance* (with Clifford W. Smith, Jr., McGraw-Hill, 1984) and *Studies in the Theory of Capital Markets* (Praeger Publishers, 1972). He founded the *Journal of Financial Economics*, one of the top two scientific journals in financial economics, in 1973, serving as managing editor from 1987 to 1997, when he became founding editor. In 1990, he was named "Scholar of the Year" by the Eastern Finance Association and one of the "Year's 25 Most Fascinating Business People" by *Fortune* magazine. He is the recipient of a 1989 McKinsey Award, the 1984 Joseph Coolidge Shaw S.J. Medal by Boston College, and was awarded (with William Meckling) the Graham and Dodd Plaque and first Leo Melamed Prize for outstanding scholarship by business school teachers from the University of Chicago's Graduate School of Business. Dr. Jensen has served as consultant and board member to various corporations, foundations, and governmental agencies, and has given expert testimony before congressional and state committees and state and federal courts. He is past president of the American Finance Association and the Western Economic Association International and a fellow of the American Finance Association, of the American Academy of Arts and Sciences, and of the European Corporate Governance Institute.

Michael A. Keller is the Ida M. Green University Librarian at Stanford University, director of academic information resources, publisher of HighWire Press, and publisher of the Stanford University Press. Formerly he has been in library leadership positions at Cornell, Berkeley, and Yale, most actively engaged in collection development with broad exposure to the global publication and bookselling trades. In 1995, in response to scholars' requests for assistance to their scholarly societies, he established the HighWire Press as an enterprise within the Stanford University Libraries to provide online copublishing services initially to three scholarly journals. As of April 2003, HighWire Press has grown to support 361 high-impact STM journals among more than 120 major scholarly societies. It is also the site creation and host service for the revolutionary, online, third edition of the *Oxford English Dictionary*. Mr. Keller is now fostering development of additional information tools and services for the scholarly community based on the successful HighWire model, such as the LOCKSS network caching application. He serves on the boards of or as an adviser to several organizations, both for-profit and not-for-profit, including the Long Now Foundation, the Digital Library Federation, the Pacific Neighborhood Coalition, and the World Economic Forum. Mr. Keller has consulted for a variety of institutions and programs, including the City of Ferrara in Italy, *Newsweek* magazine, Princeton, Cornell, Indiana and other universities, and several information technology companies as well as some of the numerous scholarly societies whose publishing enterprises HighWire Press supports. For more information see http://highwire.stanford.edu/~mkeller/.

David Lipman is the director of the National Center for Biotechnology Information (NCBI), which is a division of the National Library of Medicine within the National Institutes of Health. NCBI was created by Congress in 1988 to do basic research in computational biology and to develop computational tools, databases, and information systems for molecular biology. After medical training, Dr. Lipman joined the Mathematical Research Branch of the National Institute of Diabetes, Digestive, and Kidney Diseases as a research fellow. In his research on computational tools, he developed the most widely used methods for searching biological sequence databases. There are thousands of citations to Dr. Lipman's methods in papers, which have used them to discover biological functions for unknown sequences and that have thereby advanced the understanding of the molecular basis of human disease. Since 1989, Dr. Lipman has been the director of the NCBI, a leading research center in computational biology and one of the most heavily used sites in the world for the search and retrieval of biomedical information.

Wendy Pradt Lougee is university librarian and McKnight Presidential Professor at the University of Minnesota (appointed June 2002). As university librarian, she is responsible for a system of 16 libraries on the Twin Cities campus. Prior to her appointment at Minnesota, Ms. Lougee served as associate director of libraries at the University of Michigan, with responsibility for digital library development. Michigan's distinction as a premier digital library enterprise developed from a number of significant efforts launched during her tenure, including JSTOR, Making of America, the PEAK (Pricing Electronic Access to Knowledge) Project, OAI harvesting initiative, and a number of publisher collaborations. Ms. Lougee holds a B.A. in English from Lawrence University, an M.S. in library science from the University of Wisconsin, and an M.A. in psychology from the University of Minnesota.

Richard E. Luce is the research library director at Los Alamos National Laboratory (LANL). He was appointed project leader of LANL's Library

Without Walls in 1994, an internationally recognized pioneering large-scale digital library. Mr. Luce holds numerous advisory and consultative positions supporting digital library development and electronic publishing. In 1999 he cofounded the Open Archives Initiative to develop interoperable standards for author self-archiving systems. Currently, he is the senior adviser to the Max Planck Society's Center for Information Management, an executive board member of the National Information Standards Organization, and a member of the University of California System-wide Library and Scholarly Information Advisory Committee. He is the course director of the International Spring School on the Digital Library and E-Publishing for Science and Technology in Geneva and a founding member of the Alliance for Innovation in Science and Technology Information. Mr. Luce received a 1996 Los Alamos Distinguished Performance Award for his contributions "introducing technological innovations supporting science and technology." The research library was co-recipient of the 1999 Federal Library and Information Center of the Year award and a 1997 and 2000 Quality New Mexico "Roadrunner" recipient for organizational performance excellence based on the Malcolm Baldrige criteria.

Clifford Lynch* is the director of the Coalition for Networked Information (CNI). Prior to joining CNI in 1997, Dr. Lynch spent 18 years at the University of California Office of the President, the last 10 as director of library automation. He holds a Ph.D. in computer science from the University of California, Berkeley, and is an adjunct professor at Berkeley's School of Information Management and Systems. He is a past president of the American Society for Information Science and a fellow of the American Association for the Advancement of Science and the National Information Standards Organization. Dr. Lynch currently serves on the Internet2 Applications Council and is a member of the NRC Committee to Study Digital Archiving and the National Archives and Records Administration. He was a member of the NRC committee that published *The Digital Dilemma: Intellectual Property in the Information Infrastructure* and served on the NRC committee on Broadband Last-Mile Technology.

Jeffrey MacKie-Mason* is Arthur W. Burks Professor of Information and Computer Science and a professor of economics and public policy at the University of Michigan. He is also the founding director of the Program for Research on the Information Economy at the university and the director of doctoral studies at the School of Information at Michigan. His work is primarily in information economics, especially the Internet and advanced telecommunications technologies, and the economics of digital information content. Professor MacKie-Mason received his A.B. in environmental policy from Dartmouth, his Master of Public Policy from University of Michigan, and his Ph.D. in economics from MIT.

James J. O'Donnell is professor of classics and provost at Georgetown University. He served in 2003 as president of the American Philological Association and has been elected a fellow of the Medieval Academy of America. In 1990, he cofounded *Bryn Mawr Classical Review*, the second-oldest humanities e-journal. His 1998 book, *Avatars of the Word: From Papyrus to Cyberspace* (Harvard University Press), explores the impact of technologies of writing on the shaping of culture from antiquity to the present.

Ann Okerson* is associate university librarian for collections and technical services at Yale University. She is responsible for making digital collections available to the many and varied users at Yale and has become an expert in licensing digital information for academic use and on building consortia of libraries to achieve the most effective service at the best price for academic users. Prior to joining Yale, she served as director of the Office of Scientific and Academic Publishing at the Association of Research Libraries. Ms. Okerson was named Serials Librarian of the Year in 1993 by the American Library Association (ALA) and is also the 1999 recipient of their LITA/High Tech award. She was a member of the NRC Committee on Information Technology Strategy for the Library of Congress.

Paul Resnick is an associate professor at the University of Michigan School of Information. He previously worked as a researcher at AT&T Labs and AT&T Bell Labs, and as an assistant professor at Massachusetts Institute of Technology's Sloan School of Management. He received his master's and Ph.D. degrees in electrical engineering and computer science from MIT, and a bachelor's degree in mathematics from the University of Michigan. Professor Resnick's research focuses on sociotechnical capital, productive social relations that are enabled by the ongoing use of information and communication technology. He was a pioneer in the field of recommender systems (sometimes called collaborative filtering or social filtering). Recommender systems guide people to interesting materials based on recommendations from other people. His current research focuses on reputation systems, which apply the ideas of recommender systems to evaluating people.

Bernard Rous is the deputy director of publications and electronic publishing program director for the Association of Computing Machinery. He received his undergraduate education at Brandeis University, and an M.A. at the New School for Social Research in anthropology. Mr. Rous has worked in publishing at ACM from 1980 to the present. His responsibilities have included development and management of a database publishing system for reference publications; development of early CD-ROM and hypertext products; project manager for SGML publishing production system; development and direction of an electronic publishing program; drafting copyright and permissions policy for the networked environment; and establishment of Digital Library, Online Bibliographic Database, and Computing Portal with appropriate business models.

Edward H. Shortliffe* (*chair, Symposium Steering Committee*) is a professor and chair of the Department of Medical Informatics and deputy vice president for information technology for the Health Sciences Division of Columbia University. His research interests include medical informatics; issues related to integrated decision-support systems and their effective implementation; clinical medicine; and medical-informatics research and teaching. Prior to his current position, he was at Stanford University. Dr. Shortliffe provides expertise in both medicine and computer science. He received an A.B. from Harvard in applied mathematics, a Ph.D. from Stanford in medical information sciences, and an M.D. from Stanford. Dr. Shortliffe is a member of the Institute of Medicine (IOM), the IOM Council, and the NRC Committee on Science, Engineering, and Public Policy. He also served as chair of the NRC Committee on Enhancing the Internet for Biomedical Applications: Technical Requirements and Implementation Strategies.

Alexander Szalay is the Alumni Centennial Professor of Astronomy and professor of computer science at the Johns Hopkins University. He is a cosmologist, working on the statistical measures of the spatial distribution of galaxies and galaxy formation. He is the architect for the Science Archive of the Sloan Digital Sky Survey. Professor Szalay is project director of the National Science Foundation (NSF)-funded National Virtual Observatory.

Gordon Tibbitts is president of Blackwell Publishing, USA. Since 1982, he has worked to integrate electronic publishing technology into the traditional publishing business. Early on, he recognized the advantages of automating prepress and manufacturing work, including editorial, production, and typesetting functions, and successfully integrated the entire digital work flow at Aster Publishing Corporation in the mid-1980s. He also was responsible for automating publishing processes, creating SGML/XML/HTML-encoded content, and delivering products electronically in the early 1990s for American Health Consultants (a Thomson company). In addition to his publishing experience, Mr. Tibbitts worked for several years leading health care software development for two other Thomson information technology holdings, DKD and The Medstat Group. He developed systems sold to leading hospitals and clinics in the United States, integrating clinical and financial informatics, applying clinical best practice with performance optimization concepts drawn from evidence-based, disease-staging, case management, and JIT fields of learning. Throughout his career, Mr. Tibbitts has driven organizations toward the leading edge of technology's application in disseminating information. With Blackwell, he plays a major role in the development of all corporate technology initiatives, including Web-based content development. He has worked in executive positions for Aster Publishing, Advanstar, and The Thomson Corporation. Mr. Tibbitts received a B.S. in computer science and an M.B.A. from the University of Oregon.

Mary Waltham* is a publishing consultant. She was most recently president and publisher for *Nature* and the Nature family of journals in the United States, and formerly managing director and publisher of *The Lancet* in the United Kingdom. She founded her own consulting company 4 years ago. Its purpose is to help international scientific, technical, and medical publishers to confront the rapid change that the networked economy poses to their traditional business models, and to develop the new opportunities to build publications that deliver outstanding scientific and economic value. Ms. Waltham has worked at a senior level in science and medical publishing companies across a range of media, which include textbooks, magazines, newsletters, journals, and open learning materials. She served on the NRC Committee on Community Standards for Sharing Publication-Related Data and Materials.

Ann Wolpert became director of libraries for the Massachusetts Institute of Technology in January 1996. She oversees this distributed library system, which consists of five major collections, five smaller branch libraries in specialized subject areas, a fee-for-services group, and the Institute Archives. As of January 1999, her position expanded to include reporting responsibility for the MIT Press, which publishes approximately 200 new books and more than 40 journals per year in fields related to science and technology. Recently, Ms. Wolpert also assumed oversight of *Technology Review*, MIT's magazine of innovation. Ms. Wolpert's institute responsibilities include membership on the Committee on Copyright and Patents, the Council on Educational Technology, the Campus Plan Steering Committee, the Deans' Committee, and the President's Academic Council. She chairs the Management Board of the MIT Press, serves on the OpenCourseWare Interim Management Board, and is cochair of the Internal Review Committee for Financial Systems Services and Information Systems. Prior to joining MIT, Ms. Wolpert was executive director of library and information services at the Harvard Business School. Her experience also includes management of the Information Center of Arthur D. Little, Inc., where she additionally engaged in consulting assignments. More recent consulting assignments have taken her to Adelphi University in New York, to the campuses of INCAE in Costa Rica and Nicaragua, and to the Malaysia University of Science and Technology, Selangor, Malaysia. Ms. Wolpert is active in the professional library community. She currently serves on the Association of Research Libraries (ARL) Board of Directors and is a member of ARL's Scholarly Communication Committee and of its Internet2 Working Group. She also serves on the Board of Directors of the Boston Library Consortium. In addition, she is a member of the editorial boards of *Library & Information Science Research* and *The Journal of Library Administration*. A frequent speaker and writer, she has recently contributed papers on such topics as library service to remote library users, intellectual property management in the electronic environment, and the future of research libraries in the digital age. Ms. Wolpert serves on the Board of Trustees of Simmons College. In 1998 she was elected to the National Network for Women Leaders in Higher Education of the American Council on Education. She received a B.A. from Boston University and an M.L.S. from Simmons College.

C

Symposium Participants

Laura Abate
Himmelfarb Library, The George
   Washington University
leabate@gwu.edu

Hal Abelson
Massachusetts Institute of
   Technology
hal@mit.edu

Karen Albert
Talbot Research Library, Fox
   Chase Cancer Center
km_albert@fccc.edu

Bruce Alberts
National Academy of Sciences
balberts@nas.edu

Kerri Allen
Association of Research Libraries
kerri@arl.org

Mohammad Al-Ubaydli
National Center for
   Biotechnology Information
   (NCBI)
mo@mo.md

Carl Anderson
Drexel University Libraries
ca25@drexel.edu

Kent Anderson
New England Journal of Medicine
kanderson@nejm.org

Tom Arrison
The National Academies
tarrison@nas.edu

Susan Ashworth
Glasgow University
s.ashworth@lib.gla.ac.uk

Daniel Atkins
University of Michigan
atkins@umich.edu

John Aubry
American Museum of Natural
    History Library
jaubry@amnh.org

Midori Baer-Price
INFORMS, The Institute for
    Operations Research and the
    Management Sciences
midori.baer-price@informs.org

Marcus Banks
National Library of Medicine
BanksM@mail.nlm.nih.gov

Lori Barber
ScholarOne
lori.barber@scholarone.com

Edward Barnas
Cambridge University Press
ebarnas@cup.org

Svetla Baykoucheva
American Chemical Society (ACS)
s_baykouchev@acs.org

Malcolm Beasley
Stanford University
beasley@stanford.edu

Philippa Benson
CABS, Conservation International
p.benson@conservation.org

Sandra Berlin
American Anthropological
    Association
sberlin@aaanet.org

R. Steven Berry
The University of Chicago
berry@uchicago.edu

Iona Black
Yale University, Dept. of
    Chemistry
iona.black@yale.edu

Floyd Bloom
The Scripps Research Institute
fbloom@scripps.edu

Ronald Bluestone
Library of Congress
rblu@loc.gov

Martin Blume
The American Physical Society
blume@aps.org

Robert Bovenschulte
American Chemical Society
    Publications
rbovenschulte@acs.org

Monica Bradford
American Association for the
    Advancement of Science
    (AAAS)
mbradfor@aaas.org

Kerry Brenner
The National Academies, BLS
kbrenner@nas.edu

Sarah Brookhart
American Psychological Society
sbrookhart@psychologicalscience.org

Patrick Brown
Stanford University
pbrown@cmgm.stanford.edu

David Bruggeman
The National Academies
dbruggem@nas.edu

Edward Campion
New England Journal of Medicine
ecampion@nejm.org

Gene Carbona
The Medical Letter
gene@medicalletter.org

Bonnie Carroll
CENDI
bcarroll@infointl.com

Mary Case
Association of Research Librarians
marycase@arl.org

Meredith Cawley
AIAA
meredithc@aiaa.org

Guy Chalk
Johns Hopkins University Center
　　for Communication Programs
gchalk@jhuccp.org

Leslie Chan
University of Toronto at
　　Scarborough
chan@utsc.utoronto.ca

Nina Chang
Elsevier
n.chang@elsevier.com

Bonnie Chojnacki
University of Maryland
bc128@umail.umd.edu

Pamela Clapp Hinkle
Marine Biological Laboratory
pclapp@mbl.edu

Jeff Clark
James Madison University
clarkjc@jmu.edu

William Cohen
The Biological Bulletin, Hunter
　　College
cohen@genectr.hunter.cuny.edu

Eileen Collins
S&T Studies
el.collins@verizon.net

Jim Comes
Univ. of Massachusetts Medical
　　School
james.comes@umassmed.edu

Joan Comstock
Cadmus
comstockj@cadmus.com

Bridget Coughlin
The National Academies
bcoughli@nas.edu

C. Arleen Courtney
American Chemical Society
a_courtney@acs.org

Denise Covey
Carnegie Mellon, Hunt Library
troll@andrew.cmu.edu

Nicholas Cozzarelli
University of California, Berkeley
ncozzare@socrates.Berkeley.edu

Brian Crawford
John Wiley & Sons, Inc.
bcrawfor@wiley.com

Vicki Croft
Washington State University
croft@wsu.edu

Raym Crow
Chain Bridge Group
crow@chainbridgegroup.com

Bruce Dancik
NRC Research Press c/o
  Renewable Resources
bruce.dancik@ualberta.ca

Lloyd Davidson
Northwestern University, Mudd
  Library for Science &
  Engineering
Ldavids@northwestern.edu

Adrienne Davis
The National Academies
adavis@nas.edu

Bart De Castro
Cambridge Scientific Abstracts
bdecastro@csa.com

Laura Dean
National Center for
  Biotechnology Information
laura@idiopathic.com

Stephanie Dean
American Society for Cell Biology
  (ASCB)
sdean@ascb.org

Masako Dickinson
American Chemical Society
med96@acs.org

Heather Dittbrenner
Independent Consultant
hdittbrenner@comcast.net

Lisa Dittrich
Association of American Medical
  Colleges
lrdittrich@aamc.org

Richard Dodenhoff
American Society for
   Pharmacology and
   Experimental Therapeutics
rdodenhoff@aspet.org

Daniel Dollar
Yale University, Cushing/Whitney
   Medical Library
daniel.dollar@yale.edu

Mark Doyle
The American Physical Society
doyle@aps.org

Guy Dresser
Allen Press, Inc.
gdresser@allenpress.com

Michael Droettboom
Johns Hopkins University
mdboom@jhu.edu

James Duderstadt
University of Michigan
jjd@umich.edu

Carol Edwards
American Statistical Association
caroled@amstat.org

Anita Eisenstadt
National Science Foundation
aeisenst@nsf.gov

Julie Esanu
The National Academies, BISO
jesanu@nas.edu

Joseph Esposito
SRI Consulting
jesposito@sric.sri.com

Walter Finch
National Technical Information
   Service
wfinch@ntis.gov

Henry Flores
Copyright Clearance Center, Inc.
hflores@copyright.com

Martin Frank
American Physiological Society
mfrank@the-aps.org

Mark Frankel
AAAS
mfrankel@aaas.org

Tracie Frederick
Georgetown University Medical
   Center
tef7@georgetown.edu

Theodore Freeman
Allen Press, Inc.
tfreeman@allenpress.com

Amy Friedlander
Council on Library and
   Information Resources
amfr@loc.gov

Fred Friend
Joint Information Systems
   Committee
f.friend@ucl.ac.uk

Sherrilynne Fuller
University of Washington
sfuller@u.washington.edu

Ken Fulton
The National Academies
kfulton@nas.edu

John Gardenier
Independent Consultant
drgarden@erols.com

Lorrin Garson
American Chemical Society
lgarson@acs.org

Julia Gelfand
University of California, Irvine
    Libraries
jgelfand@uci.edu

Jane Ginsburg
Columbia University School of
    Law
ginsburg@law.columbia.edu

Carter Glass
American Geophysical Union
cglass@agu.org

Erica Goldstein
American Society for Nutritional
    Sciences
goldsteine@asns.org

Barbara Gordon
American Society for
    Biochemistry & Molecular
    Biology
bgordon@asbmb.faseb.org

Albert Greco
Fordham University
angreco@aol.com

Michael Greenberg
The Whitney Laboratory
mjgberg@aug.com

Suzanne Grefsheim
NIH Libarary
sg8d@nih.gov

Jane Griffith
National Library of Medicine
jbgriffith@nlm.nih.gov

Anne Gwynn
Library of Congress,
    Congressional Research
    Services
agwynn@crs.loc.gov

Melissa Hagemann
Open Society Institute
mhagemann@sorosny.org

Joel Hammond
BIOSIS
jkhammond@biosis.org

Charles Hancock
American Society for
    Biochemistry and Molecular
    Biology (ASBMB)
chancock@asbmb.faseb.org

Holly Harden
Welch Medical Library, Johns
    Hopkins Medical Institutions
hharden@mail.jhmi.edu

Elizabeth Have
BIOSIS
btenhave@biosis.org

Michael Held
The Rockefeller University Press
held@rockefeller.edu

Stephen Heller
National Institute of Standards
and Technology (NIST)
srheller@nist.gov

Robert Hershey
Medical Doctor
hershey@cpcug.org

Derek Hill
National Science Foundation
dhill@nsf.gov

Carol Hunter
University of Virginia
chunter@virginia.edu

Alvin Hutchinson
Smithsonian Institution Libraries,
National Zoological Park
Library
hutchinsona@si.edu

John Inglis
Cold Spring Harbor Laboratory
Press
inglis@cshl.org

Demarie Jackson
American Psychological
Association
djackson@apa.org

Beth Jacoby
University of Maryland,
Baltimore, Health Science
Library
bjacoby@hshsl.umaryland.edu

Brent Jacocks
American Speech-Language
Hearing Association (ASHA)
journals@asha.org

Michael Jensen
National Academies Press
mjensen@nas.edu

Michael Jensen
Harvard University
michael_jensen@ssrn.com

Nels Jensen
Blackwell Publishing
Nels.jensen@usa.blackwellpublishing.com

Rick Johnson
SPARC
rick@arl.org

Heather Joseph
BioOne
heather@arl.org

Lisa Junker
American Industrial Hygiene
Association
ljunker@aiha.org

Brian Kahin
University of Michigan, Ford
School of Public Policy
kahin@umich.edu

Neal Kaske
University of Maryland
nkaske@umd.edu

Cara Kaufman
Kaufman-Wills Group, LLC
ckaufman@bellatlantic.net

Irena Kavalek
U.S. Geological Survey Library
ikavalek@usgs.gov

Michael Keller
HighWire Press
michael.keller@stanford.edu

Maureen Kelly
Consultant
mckelly@ix.netcom.com

Donald King
University of Pittsburgh
dwking@pitt.edu

Gary Kittredge
Capital City Press, Inc.
gkittred@capcitypress.com

Barbara Koehler
Welch Library, Johns Hopkins
bmk@jhmi.edu

Michael Koenig
Long Island University
michael.koenig@liu.edu

Sheldon Kotzin
National Library of Medicine/
    NIH
ann_bornstein@nlm.nih.gov

Alan Kraut
American Psychological Society
akraut@psychologicalscience.org

Marc Krellenstein
Reed Elsevier
m.krellenstein@elsevier.com

Thomas Kuhn
American College of Physicians
tkuhn@acponline.org

Catherine Langrehr
Association of American
    Universities
catherine_langrehr@aau.edu

Judith LaVoie
VA Rehabilitation Research &
    Development
judith@vard.org

Ann Link
American Association of
    Immunologists - The Journal of
    Immunology
alink@ji.faseb.org

Anne Linton
George Washington University
mlbaml@gwumc.edu

David Lipman
National Institutes of Health/
    National Center for
    Biotechnology Information
dl2a@nih.gov

Wendy Lougee
University of Minnesota Library
wlougee@umn.edu

Michael Luby
Columbia University
ml1047@columbia.edu

Richard Luce
Los Alamos National Laboratory
rick.luce@lanl.gov

Richard Lucier
Dartmouth College
richard.e.lucier@dartmouth.edu

Marilyn Lux
American College of Surgeons
mlux@facs.org

Clifford Lynch
Coalition for Networked
   Information
clifford@cni.org

James MacDonald
The American Phytopathological
   Society
jdmacdonald@ucdavis.edu

Jeffrey MacKie-Mason
University of Michigan
jmm@umich.edu

Amanda Maguire
American Institute of Aeronautics
   and Astronautics
amandam@aiaa.org

Gene Major
National Aeronautics and Space
   Administration (NASA)
major@gcmd.nasa.gov

Constance Malpas
The New York Academy of
   Medicine
cmalpas@nyam.org

David Martinsen
American Chemical Society
d_martinsen@acs.org

Jan Massey
The American Association of
   Immunologists
jmassey@aai.faseb.org

Joseph Mazurkiewicz
Journal of Histochemistry and
   Cytochemistry
mazurkj@mail.amc.edu

Luke McCabe
American Institute of Aeronautics
   and Astronautics
lukem@aiaa.org

Katherine McCarter
Ecological Society of America
ksm@esa.org

Johanna McEntyre
NCBI
mcentyre@ncbi.nlm.nih.gov

Bruce McHenry
Discussion Systems
bruce@discussionsystems.com

Robert McKinney
Cadmus Communications
mckinneyr@cadmus.com

Peggy Merryman
US Geological Survey Library
mmerryma@usgs.gov

Henry Metzger
The National Academies, BISO
metzgerh@exchange.nih.gov

Cynthia Middleton
U.S. Government Printing Office
cmiddleton@gpo.gov

Kenneth Miller
American Association of Colleges
    of Pharmacy
kmiller@aacp.org

Lenne Miller
The Endocrine Society
lmiller@endo-society.org

Linda Miller
Library of Congress
lmil@loc.gov

Kurt Molholm
Defense Technical Information
    Center
kmolholm@dtic.mil

Pat Molholt
Columbia University Health
    Sciences Division
molholt@columbia.edu

Robert Molyneux
U.S. National Commission on
    Libraries and Information
    Science
bmolyneux@nclis.gov

Marguerite Montes
Journal of Rehabilitation Research
    & Development
marx@vard.org

John Muenning
University of Chicago Press
jmuenning@press.uchicago.edu

Jane Murray
University of Maryland Baltimore,
    Health Sciences & Human
    Services Library
jmurray@hshsl.umaryland.edu

Gordon Neavill
Wayne State University
aa3401@wayne.edu

Michael Neff
American Society for
    Horticultural Science
mwneff@ashs.org

Kathe Obrig
Himmelfarb Library, The George
    Washington University
mlbkso@gwumc.edu

Jack Ochs
American Chemical Society
j_ochs@acs.org

James O'Donnell
Georgetown University
jod@georgetown.edu

Ann Okerson
Yale University
ann.okerson@yale.edu

Jill O'Neill
National Federation of
    Abstracting and Information
    Services (NFAIS)
jilloneill@nfais.org

Shigeharu Ono
Kinokuniya Publications Service
ono@kinokuniya.com

David Osterbur
Harvard University
dosterbu@mcb.harvard.edu

Doris Peter
The Medical Letter, Inc.
dpeter@medicalletter.org

Walter Peter
CADMUS
peterw@cadmus.com

Stephen Phelps
Capital City Press
sphelps@capcitypress.com

Theresa Pickel
Alliance Communications Group,
    Allen Press,Inc.
tpickel@allenpress.com

Kevin Pirkey
Dartmouth Journal Services
kpirkey@dartmouthjournals.com

Barbara Pope
National Academy Press
bkline@nas.edu

Heather Price
Journal of Bone & Mineral
    Research
heather@jbmr.org

Roberta Rand
University of Miami - Rosenthal
    School of Marine &
    Atmospheric Science
rrand@rsmas.miami.edu

Alan Rapoport
National Science Foundation
arapopor@nsf.gov

Howard Ratner
Nature Publishing Group
h.ratner@natureny.com

Tovah Reis
Brown University
tovah_reis@brown.edu

Paul Resnick
University of Michigan
presnick@umich.edu

Leonard Rhine
University of Florida Health
 Science Center Libraries
lenny@library.health.ufl.edu

Susan Riedel
Library of Congress
wrie@loc.gov

Nancy Rodnan
FASEB
nrodnan@faseb.org

Linda Rosenstein
University of Pennsylvania
 Biomedical Library
rosenstl@mail.med.upenn.edu

Beth Rosner
AAAS
Rosner@aaas.org

Bernard Rous
Association for Computing
 Machinery
rous@hq.acm.org

Kevin Rowan
The National Academies,
 COSEPUP
krowan@nas.edu

Lucia Ruggiero
World Health Organization
ruglucia@paho.org

John Rumble
NIST
john.rumble@nist.gov

John Sack
Stanford University
sack@stanford.edu

Jennifer Samuels
AIAA
jens@aiaa.org

Agnes Schonbrunn
University of Texas - Houston
agnes.schonbrunn@uth.tmc.edu

Roger Schonfeld
The Andrew W. Mellon
 Foundation
rcs@mellon.org

Edward Shortliffe
Columbia University
shortliffe@dmi.columbia.edu

Elliot Siegel
National Library of Medicine
siegel@nlm.nih.gov

Pamela Sieving
National Institutes of Health
 Library
ps256e@nih.gov

William Silberg
Medscape
bsilberg@webmd.net

Natalie Simone-Fountaine
Analytical Sciences, Inc.
nfountaine@asciences.com

Robert Simoni
Department of Biological Sciences
rsimoni@asbmb.faseb.org

Susan Skomal
American Anthropological
    Association
sskomal@aaanet.org

Eric Slater
American Chemical Society
e_slater@acs.org

F. Hill Slowinski
Worthington International
hslowinski@worthingtoninternational.com

Kent Smith
National Library of Medicine
kent_smith@nlm.nih.gov

MacKenzie Smith
MIT Libraries
kenzie@mit.edu

Michael Smolens
3BillionBooks
michael@3billionbooks.com

John Sopka
National Technical Information
    Services
wfinch@ntis.gov

Beth Staehle
American Society of Plant
    Biologists
beths@aspb.org

Robert Stanley
University of Alabama,
    Birmingham
rstanley@uabmc.edu

Diane Sullenberger
The National Academies
sullenb@nas.edu

Alex Szalay
Johns Hopkins University
szalay@jhu.edu

Herbert Tabor
American Society for
    Biochemistry & Molecular
    Biology
htabor@asbmb.faseb.org

Marian Taliaferro
Association of American Medical
    Colleges
mtaliaferro@aamc.org

Patricia Thibodeau
Duke University Medical Center
thibo001@mc.duke.edu

Gordon Tibbitts
Blackwell Publishing
GTibbitts@bos.blackwellpublishing.com

Jeanne Slater Trimble
American Institute of Aeronautics
    & Astronautics
jeannet@aiaa.org

Ian Tuttle
Georgetown University
tuttlei@georgetown.edu

Paul Uhlir
The National Academies, BISO
puhlir@nas.edu

Victor Van Beuren
American Association of
    Pharmaceutical Scientists
vanbeurenv@aaps.org

Pamela Van Hine
American College of Obstetricians
    and Gynecologists
pvanhine@acog.org

John Vaughn
Association of American
    Universities
john_vaughn@aau.edu

Jack Verna
Analytical Sciences, Inc.
jverna@asciences.com

Philip Wallas
EBSCO Information Services
pwallas@ebsco.com

Mary Waltham
Independent Consultant
mary@marywaltham.com

Walt Warnick
U.S. Department of Energy
walter.warnick@science.doe.gov

Donald Waters
The Andrew W. Mellon
    Foundation
pam@mellon.org

Judy Weislogel
Elsevier
j.weislogel@elsevier.com

Nanette Welton
University of Washington
nwelton@u.washington.edu

Louise Wides
Wides Consulting, LLC
widesconsult@attglobal.net

Sophie Wilkinson
American Chemical Society
s_wilkinson@acs.org

Alma Wills
Kaufman-Wills Group, LLC
almawills12@comcast.net

Bonita Wilson
Corporation for National
    Research Initiatives (CNRI)
bwilson@cnri.reston.va.us

Nancy Winchester
American Society of Plant
    Biologists
nancyw@aspb.org

Terry Wittig
George Mason University
twittig@gmu.edu

Ann Wolpert
Massachusetts Institute of
    Technology
awolpert@mit.edu

Melissa Yorks
National Library of Medicine
yorks@nih.gov

Nevenka Zdravkovska
Georgetown University
nevenkaz@georgetown.edu

Eric Zimmerman
The Research Authority of Bar-
    Ilan University
zimmee@mail.biu.ac.il

Laura Zimmerman
Fry Communications, Inc
lzimmerm@frycomm.com